Praise for Sonny Barger's Books

On *Ridin' High, Livin' Free*

"Anyone who enjoyed Barger's first book or who would like to sit down with him over a beer will be interested in this sequel."
—*USA Today*

"Compulsive reading." —*Telegraph* (London)

"The true stuff." —*Booklist*

"Of great interest to anyone involved in the motorcycle scene."
—*Library Journal*

On *Hell's Angel*

"Maybe the baddest man on two wheels." —*Rolling Stone*

"Hell's Angel chieftain Ralph "Sonny" Barger has led one interesting, bad-to-the-bone life." —*Seattle Times*

"Sex, drugs, violence, mayhem, and a little humor. . . . A wild ride through three decades of a brotherhood that is like no other."
—*Albuquerque Journal*

"Barger's autobiography is about as subtle as a kick in the groin, and that's what's so refreshing about it."
—*San Francisco Chronicle*

"One of the best books of the year." —*Maxim*

"The hombre who gave you motorcycle culture. . . . The biker style Barger originated remains timeless." —*Chicago Tribune*

"Not your ordinary book. . . . A peek at another side of America."
—*Tulsa World*

"A ripping good read . . . as gripping a tale as you would expect from the leader of the most famous pack of all time."
—*New Times Los Angeles*

Darrell Leon for *Razor*

About the Authors

RALPH "SONNY" BARGER first burst onto the literary scene with his bestselling memoir *Hell's Angel: The Life and Times of Sonny Barger and the Hell's Angels Motorcycle Club*. He lives near Phoenix, Arizona, and is a member of the Cave Creek chapter of the Hell's Angels Motorcycle Club. Visit his website at www.sonnybarger.com.

KEITH AND KENT ZIMMERMAN are twin brothers and writing partners who previously collaborated with Sonny on his memoir. They are also the coauthors of *Rotten: No Irish, No Blacks, No Dogs,* written with the Sex Pistols' Johnny Rotten. They live in Oakland, California.

Ridin' High, Livin' Free

ALSO BY RALPH "SONNY" BARGER

Hell's Angel: The Life and Times of Sonny Barger and
the Hell's Angels Motorcycle Club

Ridin' High, Livin' Free

Hell-Raising Motorcycle Stories

RALPH "SONNY" BARGER

WITH KEITH AND KENT ZIMMERMAN

Perennial

An Imprint of HarperCollins*Publishers*

A hardcover edition of this book was published in 2002 by William Morrow, an imprint of HarperCollins Publishers.

HarperCollins books may be purchased for educational, business,
or sales promotional use. For information please write:
Special Markets Department, HarperCollins Publishers Inc.,
10 East 53rd Street, New York, NY 10022.

First Perennial edition published 2003.

Designed by Katherine Nichols

The Library of Congress has catalogued the hardcover edition as follows:
Barger, Ralph.
Ridin' high, livin' free: hell-raising motorcycle stories/by Ralph "Sonny" Barger with
Keith and Kent Zimmerman.—1st ed.
p.   cm.
ISBN 0-06-000602-1
1. Motorcycling—Anecdotes.   2. Motorcyclists—Anecdotes.   I. Zimmerman, Kent, 1953– .
II. Zimmerman, Keith.   III. Title.

GV1059.5 .B37   2002
796.7'5—dc21   2001059377

ISBN 0-06-000603-X (pbk.)

03 04 05 06 07 ❖ / RRD 10 9 8 7 6 5 4 3 2 1

To everyone

who bought and read the first book,

making this one possible,

and to those

whose stories are included.

And especially to Sarrah.

—Sonny Barger

Contents

Acknowledgments — ix

Introduction — xi

The Wandering Gypsy and the Silver Satin Kid — 1

Ed's Reunion — 17

The Further Adventures of the Peckerwood Pans — 23

The Suicide Red-and-Black-Speckled Honda — 33

MCQ3188: The Missing Years of Steve McQueen — 36

Moto Guzzi Ron's Call of the Wild — 49

Bay Bridge Discount — 58

Nan and the Basket Case — 61

Draggin' the Line with the East Bay Dragons MC — 71

Jerry's Kids — 86

Cros and Cincinnati — 93

The Saga of Repo Jim — 98

DIY/DOA — 102

Behind the Locked Door   109

When Sara Met Ron   115

The Ballad of Rocky's Green Gables   121

Blake's World-Record Whorehouse Jump   133

Loaded Linda's Silent World   137

The Ghost of Yermo   148

Laying It Down in Kayenta   154

Apple-Pickin' Time in Sebastopol   159

On the Lam   166

Take the Long Way Home   171

Shanghaied Road Tale, or It Don't Hurt 'Til the Bone Shows   177

Experts, Judges, and Covered Wagons   184

Viking Horde   188

Packin' Henry   190

Sister Teresa   196

Fitzpatrick and the Iron Chopper   200

The Last of the Unforgiven Few   207

A Coupla Paychecks fer Ya   212

Lone Wolf and the Bad Boys Crew   219

Joey Keeps His Promise   226

Kaye Keeps on Truckin'   231

Ranger Holds His Mud   235

Eli and the Raffle   241

The Cannon and the Stakeout   245

Renton's Cool Sled   248

Code of the Road   251

The Worcester Mass   258

Active Development   265

Closing Credits, Contributors, and
Unindicted Coconspirators   279

Acknowledgments

Jim Fitzgerald at the Carol Mann Agency, Fritz Clapp, Gladys Zimmerman, Deborah Zimmerman, Jennifer Sawyer Fisher, Erin Richnow, all the storytellers and bike riders who reached out, Kristen Green, Peter Grame, Scott Rourke, Chris Goff, Joe and Doris Zimmerman, Paul Butler, Melvin Shadrick, David Brunsvold, John and Leslie Adams, Murray Nixon, Freddy Falconey, Edward, Jeanne, and Tami Preciado, Lloyd Senzaki, Scott Bigelow, Road Kill, and especially Sonny Barger.

In memory of Richard Charles "Gypsy" Anderson.

—The Zimmermen

Cruising the Carefree Highway on another clear day in Arizona.

(Photograph by Jinushi, courtesy of Free & Easy magazine)

Introduction

The stories contained in this book are all motorcycle tales. All but a few involve Harley-Davidsons. There are loads of reasons to believe these stories are absolutely true. There are just as many reasons to believe they are not. This is a collection of true stories, modern myths, and biker tales, and we ain't sayin' which is which.

Our initial idea was to re-create hell-raising tales in an atmosphere of neutrality. Most of the names and places have been changed and club names (with the exception of a few) have been changed or subtly disguised. All of the stories have been streamlined to meet one important qualification: We wanted stories presented like tales a stranger sitting on the next bar stool might tell you. Truth or fiction? Who cares?

Some of the stories are serious, some funny, and some sad. Some celebrate life; others mourn death. Some feature characters who are world-famous. Some feature people who ride anonymously. Some stories glorify both outlaw and independent riders.

Some are about citizens who quietly earn a living working at a desk or running their own businesses. Some are about people who never worked a regular job in their lives. All the people we spoke with were extraordinary in their own way.

Most, if not all, of the people who submitted tales were riders. Only a few stories "nearly publishable as is" made the grade. (We weren't looking to be an outlet for creative writers.) After following up on a tale submitted to us, during the subsequent interview process we usually found details or additional stories that were even more interesting than the original submission. Since we were anxious to include a lot of motorcycle detail and escapades, people with stories were more than welcome and eager to brag a little about their rides, runs, wrecks, and mishaps.

Throughout these stories, you'll pick up on certain themes. Don't bug us about repetition. The main reasons people ride motorcycles are to clear their heads, feel the wind, and experience freedom. Those concepts are repeated throughout the book.

Not everybody's tales made it to print this time. But everyone we spoke with contributed in some way toward piecing together a cohesive but wide spectrum of motorcycle tales told by different people from different cities, towns, countries, and villages. And for that, we thank everyone who opened up his or her world to us. As a result of hundreds of interviews, phone calls, e-mails, and taped sessions, we've come up with your first batch of hell-raising motorcycle stories.

After reading these stories, do you have a hell-raising motorcycle story to share? If so, e-mail us at *story@sonnybarger.com*. Who knows? Maybe there will be a *Ridin' High, Livin' Free* sequel. Stranger things have certainly happened.

> —Sonny Barger, Keith Zimmerman, and Kent Zimmerman,
> December 7, 2001

The Wandering Gypsy and the Silver Satin Kid

Before Gypsy ever met the Silver Satin Kid, he saw his motorcycle parked outside the Continental Can Company. That's where "the Kid" worked. When Gypsy went to work there, too, he checked out the bike up close. You could tell the Kid had worked on it. There were brand-new high bars and extended chrome exhaust pipes, stuff you never really saw that much on bikes in 1957. The bike was an eighty-inch stroker from the 1940s, a Harley-Davidson with a small headlight. It didn't have a paint job. Not yet, anyway. "Silver Satin Kid" was temporarily scripted in ink on the gas tank along with a drawing of a naked girl.

The Silver Satin Kid earned his nickname from his favorite drink, Silver Satin Wine. He drank it out of a bottle in a brown paper bag when he hung out on the Oakland streets with his bike-riding buddies. To his friends, the Kid was a born leader. A big, screaming double-headed eagle—just like the one on the wine bottle label—was painted on the back of his leather jacket in silver. The eagle's sharp claws were drawn out in attack mode.

The Kid was just like that eagle—usually in attack mode. Sometimes around midnight, he'd come tearing around the corner of East 17th Street—right outside the apartment where Gypsy lived—gunning his Harley full throttle, waking up the entire east side of Oakland. The Kid usually had a pack of three or four guys from his motorcycle club right behind him, trying to keep up. They all wore matching colors on their backs and during the day they hung out at the Circle Drive-In, where their bikes took up most of the sidewalk. It would be through the Kid that Gypsy would fall in with "the Club."

Richard Charles Anderson, aka Gypsy, had wanted to ride motorcycles ever since he saw *The Wild One* back in 1954. He was seventeen, and like a lot of guys, the movie put the zap in his head. In Brando's honor, Gypsy bought a used Ariel one-cylinder English bike off a car lot in Oakland. He bought himself a cool leather shirt and wore black highway patrolman boots. He got a small tattoo on his right arm that said "El Lobo." He not only looked the part, but now he rode just like Johnny, Brando's character in the movie. He took the little one-cylinder out to the drag races. Gypsy loved that bike from day one. He'd sit on the porch and just stare at it, then jump on it and ride it around town, then park it in front of his house and look at it for a little while longer, then go riding around again. That routine could go on all night.

Gypsy hit the road impulsively for long rides alone just to clear his head. Once, on a whim, he rode from Oakland to Monterey with seventy-five cents in his pocket. When he got back, one of his

 *Watch out, Brando! Gypsy in his <u>Wild One</u> attire, 1955.*

(Photograph courtesy of Sherry Anderson)

friends remarked, "Man, you're just like a roamin' gypsy. Traveling all over and *always* alone." The name Gypsy stuck, so Richard had that name sewn onto his riding vest.

One night Gypsy was drinking at the Come-In Club with a young bike rider named Rebel. Rebel was short and skinny and didn't look much like a rebel at all. He looked more like Sal Mineo than James Dean, but he wore a V-neck T-shirt and kept the required cigarette behind his ear. Just then three Club guys came into the Come-In.

"See those three guys?" Rebel whispered. "Man, they're from the roughest, toughest motorcycle club in . . ."

As usual, Rebel kept talking, but Gypsy didn't hear much after that. Transfixed, he wanted to be just like *those* guys. He wasn't scared of them; he wanted in on their action.

As the three bikers approached the bar, Rebel called out nervously, "So, guys, what's happening?"

Walker was the leader, a lanky dude with a thin face and cold, mean eyes.

"What's happening?" Rebel repeated.

Walker ignored Rebel's small talk. "Who's your friend here?"

Rebel introduced Gypsy to the Club guys. The second fella was a dude named Crazy Cal. He was Walker's brother-in-law, not as tall as Walker, but a real stocky guy, strong as a bull, only meaner. The third member was Dakota, another serious-looking guy. Walker's first words to Gypsy went straight to the point: "What are you doing with this asshole?"

Gypsy could tell they were sizing him up. He didn't know whether they wanted to hang out and drink or kick his ass. "Come on by the Star Cafe tomorrow night," Walker said to Gypsy. "That's where we hang out, man." There was an awkward silence, then the Club guys walked off.

An old Greek owned the Star Cafe on 23rd Avenue in East Oakland. The Star Bar was right next door. The Star Bar was also where a lot of bike riders and early Club members hung out. A lot of them were ex-servicemen and former juvenile delinquents. The Star Bar had—how do you put it?—atmosphere. In blue-collar Oakland in 1957, there was a tavern like the Star Bar on nearly every street corner.

The very next day there was an empty space out in front of the Star Cafe. As Gypsy backed his bike into the curb, he noticed the Silver Satin Kid's motorcycle at the end of a long row of Harleys. The Kid had finally finished painting the frame an outrageous burnt orange with the naked girl emblazoned in yellow and black

Gypsy checks out the bikes in front of the Star Bar and Star Cafe.

(Photograph courtesy of Sherry Anderson)

two-tones. He called it the Orange Crate. Gypsy jumped off his bike and combed his hair back to make just the right entrance.

Walker, Crazy Cal, and Dakota were nowhere to be seen. Then a whistle came from the corner of the room. It was the Silver Satin Kid.

"Hey, man!" the Kid called out from the back corner. "Are you the one they call Gypsy?"

Gypsy gave the Kid the thumbs-up sign and walked toward him, nodding.

"Walker told me about you."

For being a vice president of the Club (his VP patch was stitched over the front pocket of his vest), the Kid wasn't particularly tall or sturdy. He was only nineteen, a head shorter than his fellow Club guys, with a slight, wiry build, weighing in at 155 tops. He spoke with a California drawl.

The Kid introduced Gypsy to a couple of bad-looking dudes: Johnny Slow Poke, who still had his sunglasses on even though it was well past sundown, and Tony the Wanderer, a bike-riding greaser whose hair almost hit his shoulders. Tony wasn't a Club member at the time, but wore an old, oversized khaki green army jacket. When Tony slid out of the booth to grab another beer at the bar, Gypsy noticed he had a big syringe and hypodermic needle painted on the back of his jacket.

The Kid and Gypsy shot the shit. The Kid, Gypsy learned, had never even graduated from high school. He had dropped out at sixteen, gone into the service, and worked for a time at the Granny Goose potato chip plant. Like Gypsy, he was restless, hated the cannery, hated working nights away from his buddies, and drifted from job to job. He'd saved enough money to buy gas, fix up his bike, and ride all summer with the Club. That was it. That was the Kid's life.

Soon enough, riding and the Club would become Gypsy's life as well. Gypsy began as a hang-around with the Club, riding a BSA until he traded it in for his first Harley, a 1941 Knucklehead. It wasn't long before he was hanging and riding with the Club full-time. After a few weeks, he told the Kid he wanted in. The Kid only nodded.

"Wanna get high?"

"Why not?" Gypsy shot back.

Gypsy wasn't holding anything illegal, and as it turned out, neither was the Kid. He pulled out a bottle of Romilar cough pills and spilled them out on the Formica table in his barely furnished apartment. He divided the codeine pills evenly into two piles and shoved one group Gypsy's way.

"Take all of them. Then the fun begins." They each swallowed a dozen pills, and an hour later they were hallucinating on their motorcycles. It was a cheap but colorful high. As Gypsy grabbed the handlebars of his bike, he felt as if he was wearing boxing gloves.

Joining the Club wasn't too difficult a process for Gypsy. All that stuff he had heard about needing to have a criminal record was a bunch of crap. But the Club *was* disorganized in its earliest days. It was mostly a bunch of guys who got together to get drunk. Walker, it turned out, wasn't a very gung ho president. Maybe half of the Club had bikes. Some lost them to the repo man; some sold them under pressure from their families; some just didn't have the cash, period.

The last time Gypsy saw Walker was the night he came over to see the Kid. Walker and Dakota had just gotten jobs as night watchmen. They'd prowl factories checking doorknobs. Door-rattlers. They wore uniforms and guns. To them, it was cool to wear holsters as they practiced their quick-draws in the Kid's kitchen. That night turned out to be the night Walker turned the Club over to the Silver Satin Kid.

Walker was pretty straight up about the transition.

"I'm tired of having the Club and being president. You take over, Kid. Do what you want with it. Reorganize it or make some money with it. I don't care. It's all yours." Then Walker and Dakota took off.

Guess who? The Silver Satin Kid circa 1957 on his Harley stroker.
(Sonny Barger personal collection)

Now the Kid was boss.

He took over, no problem, and in short order reorganized the Club. Pre-Kid, there weren't many rules or by-laws. Post-Kid, some members bellyached that he had gotten to be president without a vote. So the Kid threw the whole matter wide open for a vote, just in case. He was voted in unanimously. The Kid became prez—100 percent legit.

When the Kid took over as president, he practically started from scratch. First off, no girls allowed. He kept five or six of the tightest members as his inner core. The rest, if they wanted to stay, had to be re-voted in. Gypsy was member number eleven the night he was re-voted into the Club.

"How 'bout it, guys," the Kid said, presiding over the meeting. "You want Gypsy in the Club?" Gypsy was back in unanimously.

Not much more than twenty-five members were in the Club at any given time. The Kid liked it that way, a tight, loyal group—only the guys he could absolutely count on; men who wouldn't run off when it came time to stand together.

The Club in 1957 was a cool and wild lineup. Besides the Kid and Crazy Cal, there was Arnie, Rusty, Len, Merv (a cowboy from Ukiah), Joe Mendez, Swede, and two Spotted Bobs (one blond, the other with a black goatee). There were also two Als, one nicknamed Elvis. There was Ric the Blue Coffin, who had a wife called Chili Choker. There was J. C., Johnny Slow Poke, Little Dan, Toby, and Roy. Vance rode a BSA and took a ribbing for it. There was Pirate, who rode a big green hog, packing a beautiful redheaded babe, as well as Adam, Smith, Deke, and Silver. Rounding out the group was Wendell, Crew Cut, Stringer, and Gerry.

Gerry won the trophy two years running for Most Outstanding Member of the Club. Gerry was a big guy with a bigger heart. He had a black beard and slicked-back hair. He came to a meeting one night wearing a German helmet and a storm trooper's overcoat. Gerry was secretary/treasurer and also had a brother-in-law in the Club named Bruce.

In the 1950s, money was tight as hell for guys who lived to ride. Dues were only twenty-five cents per meeting, and some guys were getting kicked out because they couldn't pay *that*. Gypsy, who had a little bread, kept a bunch of guys in the Club by lending them nickels and dimes to pay their dues. The fine for fighting was five bucks, even then hardly a deterrent. Almost every week, guys would square off and end up having to feed the kitty. That kept the Club going.

The Kid's new girlfriend, Shelly, had been married to Nick the tattoo artist. His business was located right across the street from a new Club hangout called the Saints Club. One day the Kid was riding along with Gypsy when he yelled over at him, "Hey, Gypsy, you got your club tattoo yet?"

"No!"

"Wanna get one cheap?"

Kid took him to Nick's tattoo parlor.

Nick didn't have a template made yet for the new Club logo. So rather than screw around and wait, the Kid whipped out his membership card and Nick outlined the Club image from the card with an ink pen, then pressed it onto Gypsy's arm, which left a vague imprint. Then Nick got down to business, scrawling a "California" banner across the top in longhand. It looked great when it was done, with blue, red, yellow, and other colors all across Gypsy's arm. Best of all, Gypsy's Club tattoo was free because it was Nick's first. Once the rest of the guys saw it, they had to have one. But they would have to pay.

Gypsy's first wife, whom the Club nicknamed "Fraulein," really didn't mind him riding with the Club. In fact, she often hung out as well. But when she and Gypsy started arguing a lot, she ran back to her family in Montana. Gypsy couldn't live with her, but he couldn't live without her, either. He decided to try to make amends with Fraulein, so he scraped up the necessary dough for a 1,300-mile ride up to Montana. As usual, Gypsy would travel light and alone. It was late fall. Indian summer had come and gone. His plan was to jump on his spare AJS cycle, go see Fraulein in Billings, and bring her back home.

Rusty helped Gypsy roll up three blankets real tight. Then he and his bedroll were on the road. The first night Gypsy unrolled all of the blankets. But the next morning he had a bundle almost big-

Gypsy relaxes before his pilgrimage to Billings.
(Photograph courtesy of Sherry Anderson)

ger than his back wheel, so Gypsy threw two of the blankets away and folded up the last one, roughing it the rest of the way. At night he'd put his bike up on its kickstand by the side of the road, drape the blanket over himself, and sleep on the banana seat with his feet dangling over the handlebars.

On the fourth day he found himself a few miles outside of

Yellowstone Park in the pitch-black night with no place to stop. So he just kept right on riding. Then the rains came. Water came up right over his front wheel and slapped him across both sides of his face. His leathers were soaked to the bone. His boots turned into small reservoirs. When he stopped, he'd pour a quart of water out of each boot.

The weather the entire way to Billings was extreme. Daytime was so hot, large sheets of skin peeled off his arms. Nights were so cold, snowflakes bounced off him. Sometimes he'd pull over just to warm his hands by the heat of the motor.

Gypsy pulled up to a Yellowstone rest stop next to a row of garbage cans. Something big, brown, and hairy stuck out of one of the cans. As he rolled his motorcycle up closer, beeping his horn and revving his motor, a big brown bear lunged toward him. So much for resting.

Farther down the road, Gypsy got caught in a mile-long traffic snafu. He split lanes and white-lined it, dodging car doors and exasperated drivers. At the front of the snarl were four or five bears in the middle of the road begging for food from motorists. As the bears approached Gypsy, he was ready to rumble. He gunned his motorcycle louder, grabbed the chain he used to lock up his bike, and threw it over his shoulder. He made his escape swinging his chain and roaring the bike past the bears.

Unfortunately, the forces of nature wouldn't stop there. Later that night, Gypsy nearly hit a moose that was crossing the road. Gypsy was riding the center line. He looked the moose right in the eye as he shot past the large four-legged creature. Next, a black bird came out of the sky and hit him on the head, nearly knocking him cold.

By now, Gypsy had spent a grand total of nine dollars on gasoline just getting to Montana. He'd ridden so hard that by the time he hit Billings, all the rollers were worn off his chain. But his journey was to be short-lived. His attempt to woo Fraulein back was fruitless. She wanted no part of him. So Gypsy signed divorce papers at the Billings courthouse and headed back to Oakland the next day.

By 1961, life had changed for Gypsy. He now faced supporting an ex-wife and a couple of kids. He needed to find work fast. He found a decent-paying gig, but it was the dreaded night shift. Club rules stated that if you missed four meetings in a row you were automatically kicked out. Gypsy pleaded with the Kid to bend the rules; hadn't he been a loyal member for four years, from the very beginning? He tried taking a few Friday nights off, but his boss at the plant was a jerk. He threatened to fire him if he missed any more workdays. Gypsy now suffered the classic conflict: the Club or work. He reluctantly chose the job. Damn.

The last time Gypsy saw the Oakland Club was in 1967. He had been away from California for six years, since leaving the Club in '61. He was back in northern California to do some temporary construction work when he saw the Kid hanging out at the Doggie Diner in Hayward. The Kid recognized Gypsy immediately, gave him a bear hug, and the two talked for a while. Gypsy was looking to score a bag of weed to share with the guys on the site and asked the Kid if he could help him out.

"Go on down to the Saints Club," the Kid said. "I'll see if I can fix you up with somebody." Gypsy, who had just gotten paid, went down late that afternoon. When he got there, there weren't any Club members in the bar yet, just a prospect. Since Gypsy had been gone for six years, most of the current members would not know him. He was just some stranger having a half-drunk, good old time. But with each drink, Gypsy felt different: Once a Club guy, always a Club guy, he thought. As it got later into the night, the place began to fill up with locals, including a number of Club members.

Gypsy was still waiting for the Kid to show up when Pee Wee strutted into the bar. Somebody in the bar told Pee Wee about a former member sporting a faded Club tattoo. Pee Wee, seven feet and three hundred pounds of rock-solid muscle, strode over to Gypsy's table. He grabbed Gypsy's arm and raised it up for a look.

"Where'd you get that tattoo, man?"

"I used to be in the Club," Gypsy said proudly.

"When was that?"

"From the start, until '61. Ask the Kid. He'll vouch." Then Gypsy asked him, "Aren't you Pee Wee?" Gypsy remembered seeing news photos of Pee Wee getting arrested at the 1965 Vietnam protest march after fighting with the police and demonstrators.

"I suppose you think you know me or something."

"No," Gypsy said, trying to keep things cool. "I have a picture of you in my scrapbook."

Gypsy brought up the subject of weed. Could Pee Wee help arrange a buy?

Pee Wee telephoned the Kid. Meanwhile Fat Richie, a chunky Mexican about Gypsy's height, walked up to the bar.

"So you say you were in the Club? I say, so fucking what? That was then, this is now."

A hush fell over the place. Fat Richie had just put Gypsy down pretty bad in front of all the folks in the bar.

"Well, fuck you, man," Gypsy barked at Fat Richie. Big mistake. Fighting words.

Nothing happened right away, but Gypsy knew he was in a tight spot. He still needed to speak with the Kid. He wanted to grab his weed and get the hell out of the bar. Things were heating up. Gypsy walked over to Pee Wee again to ask if he could speak to the Kid himself. Just then Fat Richie turned around and sucker-punched him right in the face. Dazed, Gypsy fell back about five feet against a brick wall.

Then he saw red. With all the vengeance he could muster, Gypsy charged Fat Richie like a crazed bull. But Pee Wee stepped between Richie and Gypsy and landed another surprise punch. Pee Wee's punch ended up busting Gypsy's skull in two places, hairline fractures above and below each eye. Gypsy stumbled back and bumped into the pool table, holding his face in his hands. Whoever was shooting pool then whacked him across the back of the head with a cue stick. Gypsy slumped to the floor, broken in half.

As he rolled over to get back up, a flurry of Club fists and feet were all over him, knocking, beating, stomping, and hitting him. Three times Gypsy rose; three times he was beaten and knocked back down.

Gypsy screamed out, "I've had enough, you guys."

But the beating wouldn't stop. The stomping and hitting continued until he saw Pee Wee's size twelve boot coming down on him. It looked about three feet long and landed square on his chest, cracking a few ribs. A few inches lower and it could have ruptured his spleen.

Gypsy came within a hair of dying that night. After a few more licks the guys stopped and went back to shooting pool and BS-ing. As he got up and staggered around a bit, Fat Richie came over to him.

"What the hell did you do that for, man?" Gypsy asked, the words barely coming out.

"You know the score. You cussed out a brother," Fat Richie said.

But Fat Richie wasn't wearing his colors.

"Man," he said to Fat Richie, "I need to get back to Hayward."

Fat Richie grabbed a wet towel from the bar and took Gypsy outside to his car. Gypsy's head felt twice its size. A police car drove by. Gypsy turned away from the cops so they wouldn't see how badly he was beaten.

"Hey, man," Fat Richie asked, "you need a lift home?"

Gypsy was drunk and beaten, but he still retained a shred of pride.

"Hell, I can make it."

He got behind the wheel of his '64 Ford. He was fifteen miles from Hayward. The drive was a blur of telephone poles along the highway. He was dizzy and disoriented, but somehow he made it to his buddy's house in Hayward where he was staying.

Gypsy checked into a hospital eighteen hours later. He'd found out firsthand what it was like to be on the bad end of a Club rat pack. He was bleeding from both ears, and his face was numb. For the next six months he would cringe in pain whenever he coughed or hiccuped. Gypsy thought about getting even but decided against it. He figured he'd quit while he was still in one piece.

After that incident in the Saints Club, Gypsy never joined another club. But he kept on riding for decades before settling down for good in Oklahoma. In spite of finding himself on the wrong side of a rat pack, Gypsy remained proud of his early association with the Club.

Gypsy and the Kid were reunited in 2000, when I rolled through Oklahoma City during the *Hell's Angel* Midwest Route 66 book-signing tour. Gypsy told me that he, too, wanted one day to share his hell-raising motorcycle tales with family, friends, and the world. He liked referring to himself as "Oakland's original Gypsy." Soon after, we renewed contact and captured the story you just read.

A few weeks later, after speaking with us, Richard Charles "Gypsy" Anderson died in Tulsa on February 2, 2001, from liver and kidney complications. It seemed fitting that Gypsy's tale should open this collection of motorcycle stories. Now that his memory is preserved, from the Kid to the wandering Gypsy, I say, "Ride free, my brother. Rest in peace, my friend."

Ed's Reunion

Like most young Texans, Ed "Animal" Cargill had dreams of being a cowboy. He watched Roy Rogers every Saturday morning on the family TV. Then, bummer, his home life began to unravel. Ed's mom died before he ever got to know her. He and his new stepmother locked horns almost immediately. Ed's dad never really stood up for his son. Then one thing led to another and soon Ed's bags were packed for Cal Farley's Boys Ranch, "a Christ-centered, structured environment" where kids from preschool age through high school lived together in group homes.

In 1971, in the mind of a confused and headstrong kid like Ed, being sent to Cal Farley's was tantamount to serving time in the joint. For many young boys, Cal's was generally the last stop before reform school, juvey, or jail.

Cal Farley's was a real working cattle ranch. The boys who stayed there kept the place running. Founded back in 1939, the ranch still sits on nearly ten thousand acres of land. So vast is the Farley spread that it actually stretches from the Texas Panhandle across the New Mexico border.

On Ed's first day, he was issued three pairs of straight-legged Levi's, brogans, three white T-shirts, and a number. The only thing that separated Ed from the outside world was thirty-six miles of rocks and rattlesnakes. With no bars or fences to hold him, Ed tried several times to make his escape. His first attempt involved stealing a horse assigned to his care. He made it all the way to the New Mexico border before a suspicious rancher turned him in. Next, during a field trip to the Six Flags amusement park, Ed jumped the fence and went AWOL, living on the streets for almost an entire summer. His only mistake was visiting his father, who called the Farley Ranch and turned him in.

"When they finally brought me back, my hair had grown out all long and shaggy. I was still that same ole mixed-up kid, unwanted by my family."

Back at the Ranch, Ed was "racked," that is, confined to barracks and put on hard work detail. Ed wasn't afraid of the hard work. Hell, that was no problem. But the last straw came when he was forced to participate in the Annual Reunion Rodeo. There was something about rodeos that rubbed Ed the wrong way. They seemed barbaric and square to a kid like Ed. Besides, riding bulls hurt like hell, and he'd rather run through the rocks and rattlesnakes than ride in that goddamned rodeo again.

It was either a lack of common sense or an abundance of *cojones* that pushed Ed back out on the lam. This time he planned his escape route carefully. He'd slip out at night and head south on Highway 385 near a little town called Vega, where he would then cross over to Highway 1061 and head straight into Amarillo. By sunup the next morning, Ed would have covered a little over half the distance to Amarillo.

The Texas sun got real hot, real fast when Ed made it over the hill onto Highway 1061. He was already tired, hungry, parched, and dusty. Then came a distant rumble that grew stronger and louder the closer it came. Topping the hill was a freewheelin' pack of wild bikers. It was the deafening din of thirty Harley-Davidson choppers. Ed was amazed. Man, what a sight. Standing on the side of the road, Ed watched them roar by, one by one. Leather, denim, unshaven faces, with cool chicks leaning back on sissy bars. Their leader had a cold tombstone stare, and a cigarette dangled from the corner of his mouth. He rode front left, leading the charge, open throttle, slicing effortlessly through the dry desert air. The rest of the pack had no trouble keeping up, and they rode in a tight, wheel-to-wheel formation. Ed had seen pictures of them in the *Saturday Evening Post*. They were the world's most famous outlaw motorcycle club.

As the pack roared past Ed, the very last rider slowed down and pulled over. He motioned to Ed and pointed to the back of his bike. Ed grabbed his rucksack and climbed aboard. It was his very first ride on an outlaw chopper road machine. In a spinning cloud of dust, it was like Roy Rogers used to sing, "Happy trails to you!" The vibration of the bike tingled, numbing his legs. The ride was rigid; he could feel every bump in the road as the biker gunned his throttle to catch up with the rest of the mob. Ed said nothing, nor did the biker, and their communication was as sparse as the scrub brush that grew along the two-lane highway.

Highway 1061 looped into Amarillo, where Ed figured the authorities were lying in wait for him. Sure enough, the cops had set up a barricade up ahead, not far from the state park where a bunch of freaks and longhairs were hanging out. Two-way radios squawked. Ed didn't know if they were state or local, but any black-and-white with a bubble was trouble for him. It was plain to see: Ed was a missing kid, and the last thing these outlaw bikers needed was to draw more heat from the law for harboring some runaway.

Hopping off the chopper and high-fiving the biker, Ed made a quick run for the roadside. Hiding among the mesquite trees and cactus, he watched as the cops IDed the entire pack and detained them for over an hour. Their stoic leader seemed to keep his cool, the perpetual smoke still dangling from his lips. Finally the pack was waved on and escorted down into the park area.

That night, Ed slipped past the barricade and down into the grounds. But he kept his distance. The park had erupted into party central. Ed could smell the burning reefer and the barbecued steaks. The bikers were having a great time, horsing around and fighting, drinking beer, some reclining on their bikes. The cops also remained in the distance, vigilant and watching. Ed crashed that night in the park as images of bikers danced in his head.

Ed stayed out for another week, living rough, eating and sleeping wherever he could before he was eventually caught and hauled back in. Back at the Ranch, Ed was now shoveling horseshit and going to school. He couldn't decide which he hated more, the horse stalls or the classroom. For one writing assignment Ed wrote about how one day he would become a biker, just like the guy who picked him up out on Highway 1061. Someday, he wrote, he would be somebody, ride a motorcycle and take shit from no one.

None of this sat particularly well with Ed's teacher. As punishment, Ed was told to stand up and read his essay in front of the class. What was supposed to humiliate Ed only fueled his pride and gave him a goal, something to strive for. Nobody could erase those biker images that were burned into Ed's brain, not even the Ranch headshrinker. Then Ed's running days were over.

"Hang in there," he said to himself. "Just one more year."

Ed decided he would stick out school and soon be his own man.

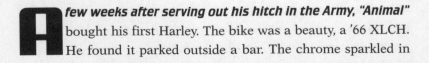

A *few weeks after serving out his hitch in the Army, "Animal"* bought his first Harley. The bike was a beauty, a '66 XLCH. He found it parked outside a bar. The chrome sparkled in the setting sun. Animal strode inside the tavern and asked the guy chalking up his pool cue, "Hey, partner, is that your bike?"

After a lot of haggling, a few pitchers of beer, and a couple games of pool, the bike was his for the $1,800 in military back pay Animal had stashed in the front pocket of his Frisco jeans. He handed over the cash and took the chopper out for a test spin. It rode as great as it looked. Animal turned the bike around and pulled back up to the entrance of the tavern for the pink slip. Then he rode off to begin his calling as a two-wheeling, 1%er outlaw biker.

Animal felt uneasy at the reunion. To say he stuck out of the crowd would be an understatement. His long hair fell way past his shoulders; his buffed-out arms were covered in club tattoos. He wore his colors proudly, a president patch stitched onto the front of his vest. It was strange to be coming back, especially to the annual Reunion Rodeo at Cal Farley's Boys Ranch.

"Ed, is that you?"

It had been a long time since anyone had called him Ed. Animal's presence definitely tripped out some of the old-timers.

Coming back for the reunion exorcized a few demons from Animal's past. The Ranch had changed a lot since the days he lived there. Farley's Ranch seemed a lot less isolated from the outside world than he remembered. Now, five hundred kids lived productively on the premises. Gone were the mean-faced dorm parents; trained counselors had replaced them a while back. No more "racking" and other such punishment. Schooling now took precedence over Ranch chores. Instead of forty scruffy boys warehoused in one dorm block, there were four boys or four girls assigned to each room.

Girls? Cal Farley's *Boys* Ranch was now coeducational!

Revisiting the Ranch reeled Animal back to his adolescent days, when authority pushed him hard and he pushed back harder.

"Back then, I never trusted anyone over thirty," Animal recalled.

Animal, a Cal Farley's Boys Ranch alumnus, is a proud 1%er prez.
(Photograph courtesy of Ed "Animal" Cargill)

"It was a tough place. We were surrounded by tobacco-chewing red-necks."

But Animal is hardly the Cal Farley poster boy for conformity. More than anything, it was that pack of wild, roving bikers that rescued him from the desert that made Animal what he is today—proud of his club, his brothers, his old lady, his bike, and his patch.

As he strolled the grounds of Cal's one last time before riding off, he wondered which kid inside those walls was as confused and full of spite as he had been thirty years earlier. Then Animal laughed at the irony of it all. What a trip. From outside appearances, he was the guy who had changed the most over the years. But on the inside, Animal was the one who had changed the least.

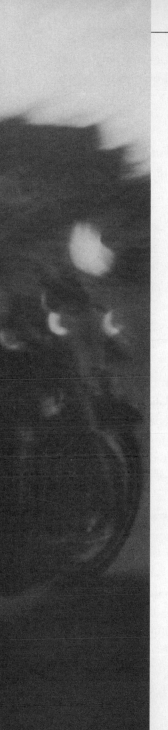

The Further Adventures of the Peckerwood Pans

Cincinnati got out of the joint in February 1977 after doing a lengthy term in Folsom Prison. He was one happy guy when Folsom saw the back of him. All Cincinnati could think about was being with his brothers on the street, riding, riding, and riding some more. But there was no way he could have been prepared for his first USA Run.

For some of us Club members still in the Folsom pen during the mid-seventies, the USA Runs were brand-new, a tradition born while we were locked away. It was an annual run for members from the entire country. Held in a different location each year, it was mandatory for all Club members to attend . . . unless they had a really good excuse. At first we assumed it wouldn't be much different than a putt from Oakland to Berdoo, only further. Well, we were wrong. It was a hell of a lot more than just a ride.

Cincinnati's best buddy, Delbert, rode around to the flat where Cincinnati was staying, and yelled out to him from the street, "I'm game to run. So, you on?"

"On?"

"The USA Run."

Since his release, Cincinnati had been in the middle of moving into new digs. Unfortunately, the start of the run interfered with his move-in date. Bad timing. But Cincinnati had a plan.

"Tell you what, Delbert. I'll ride along if you can give me a few extra days to get my shit moved in."

Delbert shrugged in the affirmative. Sure. No problem. He'd wait.

But Cincinnati couldn't wait. Come moving day, Cincinnati couldn't ignore "the call of the run," so he ended up just stacking his shit in a big pile in a corner of his new front room, and got a friend to take care of the rest until he got back. Meanwhile, a couple of Club riding brothers Dru and Bennie decided they, too, would ride along with Cincinnati and Delbert. All four guys were fresh out of the joint, and none of them had ever been on a USA Run, either. They all rode rigid-frame choppers running Sportster tanks, straight pipes, and a lot of front end.

Preacher is a forty-year member, and one of Cincinnati's closest friends in the Club. He knew the score when it came to the ins and outs of surviving a Club run and coming back feeling good. Preacher tried to give "the rookies" the game on running cross-country, but apparently the four Folsom grads were not listening. His advice was sensible and right on the money: When you get ready to call it a day, it should be early enough to find a good place to camp. And *always* gas up the night before. That way, when you get up in the A.M., you can pack up, put on some A.M. miles, and have breakfast at the first gas stop. Otherwise, Preacher warned, you'll never get on the road and you'll never make it to the final destination, Yankton, South Dakota. It was all good advice, and the four should have listened.

Of course, all four met late at the Oakland clubhouse, grabbed a picture of yours truly (I was still holed up in Folsom in 1977), and finally hit the road. That first day, they got as far as Truckee, a town near the Cal/Neva border. They could have ridden further, but they got a little carried away drinking and fighting in a bar and, well, you know how that all goes. Over shot glasses and pool cues, the quartet pledged that they would write me every day. Since three out of the four were running Panhead cases with Shovelhead tops, they would call their trip (and my letters to Folsom) the Further Adventures of the Peckerwood Pans. It was Peckerwood, as in prison slang for a dirty white boy, and Pans because they rode Panheads.

They would start each letter with a dateline like "Today we find our heroes in a bar in Truckee." Then they would proceed to recount the day's journey—the ups, downs, and details of their very first USA Run. Believe me, each letter was almost like riding along with them. Every day at Folsom, all my guys got so wrapped up in their adventures, they couldn't wait for me to hit the yard with the latest installment.

Heading out of Truckee and across the Nevada border, the four didn't know that Nevada had recently passed a helmet law. By the time Cincinnati, Del, Dru, and Bennie got to the middle of the state, they were spotted and pulled over by the cops. Of course the lawmen didn't believe Cincinnati when he claimed they hadn't heard of such a stupid law. And when they figured out that all four were fresh out of the pen, that didn't help much, either. Ignorance of the law wouldn't wash.

Fortunately some Joe Do-Gooder came to their rescue. Bennie jumped into his pickup, rode into Winnemucca, and came back with some lids. The four got as far as Wendover before calling it a night. This would turn out to be the only night the Peckerwood

Pans attempted to camp out. From that night on, it was indoor plumbing, motels, and night running, since these guys never made it on the road earlier than three or four in the afternoon.

Somewhere into Utah, Dru developed a case of knock-a-noia, and was convinced he had peeled the hardface off his cam lobes. He just couldn't keep his dick skinners off his push rods, so he had them adjusted so far out that his rocker arms were banging into his shovels.

The four stopped at a Harley shop in Salt Lake City and got some parts, where they ran into a probate for a local B-grade club (ain't 1%er, ain't AMA—American Motorcycle Association) and followed him to his clubhouse. As they all went inside, the phone was ringing, so the probate ran to the back to answer it. Delbert went behind the bar to check out the inventory and the probate screamed and hollered at him, pushing Big Del (or at least trying to) out from behind the bar. Delbert hit him and tossed him out the window into a neighbor's yard. All the chump had to do was say they had a policy of nobody but members behind the plank and there would have been no problem. But no, they had to resort to the rough stuff.

Delbert carried the probate back inside and had him call his prez. They could hear him over the phone, prez telling probate to lock up the fucking place and split because he'd heard that the Club was in town (word travels fast). The probate promptly lost it.

"Too late, asshole," he gasped, "they're already here. I've been beaten silly and tossed out the window, and if the rest of you boys don't get down here quick-like, I'm giving up the keys and splitting."

The rest of their club showed up and Dru got most of them crowded around his bike. They were all holding lamps, lightbulbs, and anything else that would give off light. There were more extension cords running off the house than you could imagine. Dru was sitting on the ground talking shit, his gear cover off, using his fuel line to wash parts.

"You guys ain't nothing but a bunch of fucking jerks," spat Dru.

"You was jerks when we got here and you'll be jerks when we leave. Who's got a whetstone? I need to sharpen my Gerber."

And shit like that.

Dru had a girl on the other side of his bike with a mirror propped up against his bike. She was brushing her hair. Bennie was sitting off to the side, kind of relaxing. Dru told this lady to do something and she told him to wait. "Can't you see I'm busy?"

Bennie let out a scream and came running over, picked up a case of oil, and banged it across the top of her head. She was lying on the ground moaning and Dru said, "I told you someone was going to get tired of your shit, so quit your whining and do what I say."

This really got the other guys nervous. You could see their shadows trembling. They had this one guy, about six-six or six-seven, named Big Bob, walking around with a tag on his patch that read "Enforcer." Cincinnati had never seen such a tag on anyone before and asked him about it. But apparently enforcers aren't allowed to talk to the lowlifes. He just walked away. Prez had been sucking on a bottle of whiskey ever since he got to the clubhouse and was starting to get drunk. He had been advising Dru on the fine art of motorcycle maintenance since he'd arrived. Seems he ran a shop in town. All of a sudden Dru growls, "Oh, shit."

It got real quiet. Delbert took stock of the situation and told their prez that they needed to take a walk. Prez's advice had just knocked Dru's cam bearings into the cases. Delbert and the prez walked around the front of the house. Then came the sounds, *smack*, *smack*, *smack*, and this guy starts begging. Cincinnati ran around, with the rest of their mob behind him, and Dru behind them. Delbert was beating the bark off this fool.

"One on one," someone hollers.

Cincinnati spun around, put his hand on the nearest guy's chest, and shoved him back into the crowd.

"That's right, motherfucker," Cincinnati yelled into his face, "one on one, and my guys against your guys."

Someone started up a chant of "Big Bob, Big Bob."

Cincinnati was shoving guys backward and Dru was stabbing them in their asses from behind. Meanwhile Big Bob was slowly squatting down; this chump went from six-six to four-five in about three heartbeats.

"Oh God, no," the club said.

Delbert grabbed Little Bob (formerly Big Bob) out of the crowd and told him he was in charge. Bob wasn't exactly happy with this new deal. Delbert and Cincinnati sat him down and convinced him that he needed to salvage a bit of his self-respect.

"Just listen to us, do what we tell ya, and you will be a hero in the eyes of your mob when we leave."

They got Dru to a motel and found the keys to the ex-prez's bike shop. A couple of guys pulled Dru's engine while Cincinnati and Delbert "shopped around" the room. Sitting on the bench was a brand-new fresh 96 slab side shovelhead engine. It had just been delivered from a shop in Ogden. Delbert actually knew the guy from Ogden from someplace or another—Delbert knew everybody—and called him up. The guy was an excellent wrench and thoroughly guaranteed the engine. Come morning, Dru's engine was in a crate headed back to Oakland, and the new engine was mounted on his bike. Dru wasn't happy with the Utah registration and license plate, but he lived with it.

Throughout the whole ordeal, Cincinnati had been calling the run site. The run was located at the Lewis and Clark campground near Yankton, South Dakota. Getting further and further behind, they were worried about even making it, but New York City told them that no matter what, they would be there when the Four got there. Finally the Pans hit Highway 81 in Nebraska and turned north. All the way up the road they kept passing club guys going the opposite way, heading home. They waved like motherfuckers, but the closer they got to Yankton, the fewer and fewer guys they saw. They figured by the time they hit Lewis and Clark, no one would be there.

The Four rolled into the run site about 11:30 P.M., and sure enough, New York City was there, along with another bunch of hard-core motherfuckers, all waiting to party. The Peckerwood Panheads were happy as hell. They pitched their tent in twelve minutes flat, barely making it, because by midnight the run was officially over. Next morning, everyone who had waited in Yankton headed to Omaha. By the time the guys rolled into the Omaha clubhouse, multitasking Baz, a good friend and outstanding rider, had a phone jammed between his shoulder and ear, had steaks on the barbecue, was rounding up the bitches to get a bunch over for the fellas, *and* was whipping up a cake in a bowl he was holding. Get on, Bazwell!

They had been in Omaha a few days when they got a phone call from Harry from New York City. He was in Custer, South Dakota, at some kind of German inn. He said it was ultra-cool there, that a couple of guests had been showing him all their old Nazi ID. Harry and his old lady were both Jews.

"Grab some of the guys, ride up, and join us!"

Seemed like a good idea to about twenty of the boys. They were in Custer in a flash, spent the night, and the next day were out and about checking the terrain. They checked out the place where the Chief Crazy Horse Memorial was just starting to be carved. In 1977, the memorial was only a hole in the mountain, and not much more. They met up with the guy doing the carving, made the right arrangements, and the following morning he took the mob on a private tour. From the highway looking up, the monument didn't really look all that big. But up in the hole was a whole different story. Flat fucking impressive.

In a bar in Custer, Cincinnati and the guys met a biker from the Sock Fuckers MC in Montana. Whether or not they still exist, who knows? But it's a ballsy name, whether they were serious or just fucking around. The crew then left Custer, and after a long night in Deadwood, South Dakota, they rolled into Montana. They stopped for gas in the middle of one of the reservations and, after gassing up, looked for a place to chow down. They were politely directed to a joint up the street, and all rode in resonant unison. The eatery must have been some kind of Indian teen club or something. The guys stood around eating and shooting pool with the Indian kids. Some of the Club guys took the kids outside to show them the bikes. They'd been there an hour when someone ran up to Cincinnati and told him that maybe he should go outside and check up on things. Cincinnati walked out, and holy shit!

There must have been over a hundred Indians standing up and down both sides of the street. Not kids, but adults. Big-time chiefs. Medicine men. Warriors. Cincinnati sent someone back inside to alert the crew to start saddling up. The vibe was getting heavy. One gentleman (the chief?) stood tall, in front of all the others. There was something about the way he stood, and the way the other folks acted toward him; he seemed like The Man.

Cincinnati rode across the street, spun a U-turn, and stopped right beside The Man. He didn't move a muscle. Delbert gave Cincinnati the sign that the troops were ready to roll. Cincinnati looked up at this personified Indian head nickel. Their eyes briefly met, then The Man smiled real fucking cool-like. He knew, and he knew that the boys knew that he knew. Real cool vibrations were in the air. Cincinnati rode to the front, next to Delbert, and it was wagons ho. Cincinnati and the guys had treated their kids like human beings, and the warriors flat-out did not blame any of them for the sins of their fathers.

Later that night, the party pulled into a little one-bay Texaco gas station right next to the Custer battlefield. It was raining hard and so dark out that their headlights were totally useless. The only time you saw the road was when lightning flashed, and the flashes were damn huge. Everybody got gas and asked if there was someplace to get rooms.

The Indian kid running the station placed a call.

"The only motel anywhere is filled."

The kid and a couple of his friends had a van in the bay they were trying to fix up. A few of the guys were giving them some pointers when the phone rang again. The kid answered and hung up.

"There's rooms at the motel, and here's how you get there."

Some of the guests had just been told to pack their shit and get out. When Cincinnati, Delbert, and the guys got to the office, no one was there, just a note and a pile of linens. The note instructed them to make their own beds, that there would be no charge, and that the owners hoped they'd be comfortable. Apparently one of their horses had been hit by lightning, and they were out looking for it in the storm.

The next morning, the sky was so blue it hurt to look at it, and sitting outside every biker's door was a watermelon. They never did figure out the meaning of the gift. Cincinnati and the guys toured the Custer battlefield with an outlook that no other group of white people had ever toured the place with. Bottom line: Custer was not only a fool but also a fool who had fucked up.

Rolling into Yellowstone Park from the northern entrance on a road called Highway of the Gods, the bikes barely made it, the altitude was so intense. Harleys don't like running without air. Rooms were rounded up at Mammoth Hot Springs as the boys hit the bar and had a ball. The next day while riding through the park, they stampeded herds of buffalo, which scared the shit out of the tourists.

The day after that they went around a bend coming into Rexburg, Idaho, into a large garage that had a big sign that read GRIM REAPERS MC. Cincinnati had known a guy in the pen called Reefer Jim and knew that he'd just moved out of California. They stopped in, and, yeah, it was the same dude. They wined and dined for a couple of days, performed the necessary bike repairs, and then it was back out on the road. A few days later, the Peckerwoods would catch up with the Labor Day Run that was being held on Highway 395 in California.

The original four who had left Oakland rode back to the Bay Area together. As they neared the Caldecott Tunnel that empties into

Oakland, a carload of idiots did something that required a little chasing down. After a short but frank discussion, it was off to the clubhouse to write the final chapter of the Further Adventures of the Peckerwood Pans.

Several weeks later, Sharon, my old lady at the time, stopped by Cincinnati's new place and asked him about the last chapter of their adventures. What the hell happened? The guys on the yard at Folsom were left hanging.

After tearing the house apart, wondering what the fuck, Cincinnati found the final installment in the back pocket of a pair of pants in a pile of dirty clothes. The letter was all greasy, a little worse for wear and tear, but intact. My men at Folsom finally got the whole story and made it through another day in the joint, thanks to the Further Adventures of the Peckerwood Pans.

The Suicide Red-and-Black-Speckled Honda

Nic Tolbert was determined to ride to Bass Lake. The night before we left for our annual Bass Lake Run, Nic totaled his bike outrunning the cops. Well, almost. He'd been out on the Ave (Telegraph Avenue) in Berkeley, California, trying to get lucky with one of the college bitches when he attracted the attention of the law. Not needing yet another ticket, and especially not wanting to see a cop because of several outstanding warrants, he took off, with the Berkeley fuzz following in hot pursuit.

Nic would have made his escape except for smacking a Volvo (the official NASCAR vehicle of Berkeley) that pulled out in front of him. Getting himself and his bike out of jail exhausted nearly all of his and a few girlfriends' time and money. His bike was in somber shape, and as our 6:00 A.M. leave time neared, there weren't many of us available to help Nic get his ride in runworthy shape. Nic was out of time, money, and parts.

Then . . . he made it!

At six in the morning, Nic met the Club at the bar where we were scheduled to leave. That was the good news. The bad news was that he was riding—ahem—a red Honda. Now, we didn't know if he had knocked someone off of it with a two-by-four (Nic was that desperate to make the run), stole it off the streets or out of somebody's garage, or just borrowed it (doubt it). While he lacked the cycle-stealing skills of some of the better-known East Bay bike thieves, it seemed a sure bet Nic didn't buy said Honda. A sack full of spray paint cans made us further doubt the validity of his acquiring the bike legitimately, which didn't stop Nic and the new anti-Honda guys from manning rattle cans of flat black paint and making quick work of Nic's little red darling. Whatever happened, Nic made the run and avoided a stiff fine from the Club.

In the fab 1960s, our gas stops were a skimpy fifty or sixty miles apart. Because we all ran on small tanks, a fast bike would not go much further than sixty-five miles on a tank full of leaded. After a couple of stops and tangling with the anti-Honda guys in the Club, Nic's spray can work, meant to cover up the original Honda red, was pretty much in vain. By the time we got to the lake, he had had so many bloody noses that the Honda was almost red again.

I guess he finally took one punch too many. Once we rolled onto the run site, shut down our bikes, and started climbing off, we heard a bloodcurdling scream and then a noise we had never heard before on God's green earth. We gandered toward the trees and saw a huge cloud of dust coming straight for us. The scream came from Nic and his suicide red-and-black-speckled Honda. He had jerked the exhaust system off of the Honda and was going about as fast as the Jap piece of shit would go . . . right down the middle of where seventy or eighty of us were standing.

As the crowd quickly parted, Nic headed straight for the lake, by way of a twenty-five-foot cliff.

SPLASH!

He landed right out in the middle of a cove. We could see his head pop up about fifty yards or so out in the water. A cheer went up around the whole Club, which was cool. What wasn't cool was that, like me, Nic couldn't swim a stroke. He was out in the deep end yelling for his life.

"Help . . . save me, help, help!"

Everybody thought he was fucking around as usual. Some of the guys even tossed rocks, beer cans, and other shit at him as he flailed for his life in the water. Slowly, when we began to realize that Nic just might be in trouble, we saw this big-assed water ski boat come idling by. It had a full-blown 454 engine in it, and a blond, tanned, typical Southern California creep at the wheel with six or seven bikini-clad bitches hanging all over him.

They plucked Nic out of the water and started to turn the vessel toward us. Then Nic whispered something into the driver's ear. The boat stopped, turned back around, and Blondie hit the gas as they took off in the opposite direction for the open water.

Nic looked back at us and gave us all the finger. Then he dived headfirst into a sea of titties. We figured those babes agreed; you *do* meet the nicest people on a Honda.

MCQ3188: The Missing Years of Steve McQueen

Today, it's commonplace for Hollywood movie stars to ride motorcycles. To be seen or photographed on
a motorcycle is now considered
macho for the men, sexy to the women. But
during the 1960s and 1970s, one actor, Steve
McQueen, took that concept to the forefront.
He helped advance motorcycling as a sport
when it was considered merely risky and daredevil antics. Even before his acting career took
off, he rode and raced, earning a reputation as
a serious competitor and scrambler, making
noise on both the off-road trails and enduro
circuits. In 1966, he tested and reviewed a slew
of dirt and street motorcycles for *Popular Science* magazine. In 1971, he was featured on the
cover of *Sports Illustrated*, promoting desert

A Steve McQueen the public rarely saw...Indian T-shirt and a face full of character.
(Photograph by Barbara McQueen Brunsvold)

scrambling as a sport. In the years leading up to his death in 1980, McQueen's love of machinery eventually overtook even his interest in making films.

After collecting a share of the profits from the 1974 disaster film *The Towering Inferno*, Steve McQueen became the highest-paid movie star in the world. Shrewdly grabbing a percentage of the film's receipts, Steve took home almost fourteen million bucks, an astronomical sum for the 1970s. At about that same time, McQueen wanted to break away from the plastic Hollywood lifestyle. He yearned to expand his mind and simplify his life. By the mid-1970s, Steve McQueen was in personal exploration mode. His love for motor vehicles played a big part in his "great escape" from the

Hollywood spotlight. By 1976, McQueen was forty-six years old and on his way toward dedicating the remaining years of his life to collecting and restoring all types of motorcycles, cars, and (eventually) airplanes.

At the same time, a young model named Barbara Minty became one of the top cover girls in America. Barbara grew up on an Oregon dairy farm. As a teenager she hopped on a plane to New York City and became a model for the Ford Agency. In addition to fashion shows, Barbara appeared on numerous magazine covers such as *Cosmopolitan* and *Vogue* and graced the pages of *Sports Illustrated*'s swimsuit editions.

One day in 1976 Barbara was relaxing in Idaho when she got a call from her agent. Could she possibly come to Los Angeles to discuss a part in a movie? Steve McQueen was interested in casting her as an Indian bride for one of his upcoming films.

"I was twenty-four, successful, and doing really well," Barbara remembered. "I went to Los Angeles. All I knew was that I was meeting Steve McQueen, the famous actor. But the guy who answered the door was shaggy and short. He wasn't heavy, though he was built. I had no idea he looked like that. I guess I was expecting Paul Newman. I probably said about two words while he just sat there, stared at me, and drank his beer with ice cubes. Then I left."

After the meeting, Barbara turned to her agent and said, "There's something about that guy. My God, I just love that man."

Her agent was perplexed. "How can you *love* him? You only said two words to him."

The following day by the pool at the hotel, McQueen approached her and introduced himself again.

"Mind if I buy you a beer?"

As the two got talking, oddly, McQueen walked away in midsentence. Barbara dismissed him as an eccentric Hollywood type and went back to her reading. But soon Steve came back, took Barbara by the arm into the sauna, shut the door, and began asking her a barrage of personal questions.

"These were serious questions. What were my beliefs and thoughts on certain issues? What kind of values did I have? He was asking the kind of questions a father might ask a young man before blessing the marriage of his daughter."

McQueen arranged for her to have dinner that night at his place. Barbara was tentative but agreed.

"I rang the bell and he opened the door," Barbara recalled, "and there sat two blondes on the couch. When I arrived—his little princess—he rushed the girls out. We took a drive and didn't come back for a day or so, and that's how it happened."

Over the next several months—he in Los Angeles, she in New York—the two would meet in Denver every other weekend. He loved the effect beer had on him in the mile-high altitudes of Colorado. If they met in Idaho, McQueen would sometimes drive the whole way from California, hopped up on "vitamin C" pills. The two would continue on up to Montana, take in the scenery, and scout out land and ranch houses. It wasn't long before Barbara decided to leave New York. It was time to move to California to see where this strange relationship would take her.

The Steve McQueen Barbara fell for was the product of a rough-and-tumble childhood. His father flew airplanes for a while and left his family when Steve was quite young. McQueen's mother wasn't exactly stable, either. She placed young Steve in a reformatory, the California Junior Boys Republic in Chino, which assigned him ID# MCQ3188. The accounts of his life are inconsistent. According to one biography (*Portrait of an American Rebel*), at age sixteen he joined the Merchant Marine but jumped ship in Port Arthur, Texas, and took a job as a towel boy in a brothel. McQueen drifted about, working for a spell as a lumberjack, taxi driver, sandal maker, and boxer. By 1952, at age twenty-two, he joined the Marine Corps and immediately butted heads with authority. After stretching a weekend pass into two weeks AWOL, he did forty-one days in the brig. But McQueen eventually gained an honorable discharge in 1954. By the mid-fifties, through a girlfriend, he got interested in acting. After a couple years of study, he was invited to join New York's prestigious Actors Studio, one of only two accepted from two thousand applicants.

Before his movie career took off, McQueen became one of the very first television superstars, playing Western bounty hunter Josh Randall on the hit series *Wanted: Dead or Alive*. McQueen starred in several hit movies, including *The Magnificent Seven, The Great Escape, Bullitt, The Thomas Crown Affair, The Getaway, Papillon,* and *The Sand Pebbles,* for which he was nominated for an Academy Award.

Steve McQueen became one of the first stars in Hollywood to command over a million dollars per picture. Yet when he met Barbara he was at a serious spiritual juncture in his life. McQueen craved stability and peace of mind. He was at a point in his film career where he could have made more movies and more money, but chose a less intense schedule. He wanted a stay-at-home mate and a traditional home life, the kind he had never enjoyed during his youth. Steve and Barbara moved in together in '76, a short time after he "auditioned" her in the sauna.

Appearance-wise, McQueen was entering his funky, relaxed stage. His long, thick, curly hair, wire-rimmed glasses, and scraggly beard kept him from being recognized on the streets. Not long after Barbara moved into his Malibu home, McQueen came home with a pair of overalls and an old Harley basket case.

"Barbara," he said, "today we're going to put a motorcycle together."

McQueen's garage was filled with a variety of bikes and parts. One of his most prized possessions sat in his living room—the Red Bike, an early-twentieth-century Indian Board Track Racer. Over

McQueen aboard one of his many meticulously restored Indians.
(Photograph by Barbara McQueen Brunsvold)

the years, that bike would become a monument in every living room in every home McQueen lived in until his death.

Barbara's first motorcycle ride with McQueen was also her first experience on the back of a fast cycle. McQueen's rat bike, like most rat bikes, was a funky, dirty collection of stray parts and personality. The McQueen rat bike started out as a 1948 Indian Chief before ultimately being chopped. With Barbara holding tight on the back, the two ripped down Kanan Dune Road, swung around to the San Diego Freeway, then back down to the Pacific Coast Highway.

Before heading back to Malibu, the couple stopped off at the famed Rock House for a couple of beers.

"I had tears running down my face from crying," said Barbara, remembering the rat bike ride. "While I was sure he was a good rider, he *was* being a little reckless. I was scared to death. Looking back, it might have been cool riding with Steve McQueen, but at the time it was 'Help! Take me home, I never want to see you again!' Actually I think he was pushing my envelope to see how much I could endure. He was always doing that. That was his style."

McQueen continued to push Barbara's conventional limits. One day he came home and announced, "Honey, I just bought a hangar and an airplane."

McQueen pedaling an antique Harley outside "the Hangar," Steve and Barbara's home for about two years. (Photograph by Barbara McQueen Brunsvold)

Santa Paula was located ten miles inland from Ventura, not far from where McQueen and his friend Sam "Mr. Indian" Pierce restored McQueen's large collection of antique Indians, Harleys, and Aces. He had rediscovered the small town and, yes, bought the airplane hangar at the little Santa Paula Airport. "The Hangar" became Steve and Barbara's new home.

"We lived in the Hangar for almost two years and it was great," said Barbara. "The Hangar became his club. Inside were two airplanes and maybe twenty-five bikes. We had a bed in one corner and a small kitchen in the other. I can't remember where the shower was, but if you wanted to go to the bathroom, you stumbled outdoors with the keys. We finally put a light inside the Hangar because every time I got up in the middle of the night, it was so dark I'd get jabbed in the guts with a pair of motorcycle handlebars."

Life in the Hangar with McQueen went incredibly smoothly. The actor-turned-Bohemian-grease-monkey was finally in his element. Many of the airport locals felt welcome enough to hang out while McQueen swilled beer and ice cubes in his stocking feet. Each morning, lying in bed with her morning coffee, Barbara could flick a switch and the whole front of the hangar would open up, revealing an expanse of airplane runways and sunny hillsides. A steady stream of bikers, mechanics, and pilots would drop by to drink, putter on the workbench, or just plain shoot the bull.

The two would frequently take long, leisurely country motorcycle rides. Or else McQueen would hit the off-road to play around a bit. But by this time, as a rider, he'd mellowed his act considerably (except, of course, when he rode the rat bike). With Barbara and a couple of close pals in tow, McQueen scoured antique bike shows and motorcycle swap meets, cruising for old motorcycle parts in order to finish the numerous bike restoration projects he and his friend Mr. Indian undertook.

Every accessory in McQueen's new life was either old or rebuilt—motorcycles, cars, furniture, appliances, and assorted an-

tique knickknacks like gas pumps and scales. The only modern convenience Barbara insisted on was a washer and dryer.

"The washer, dryer, and I were the only things under fifty years
old that were part of his life."

While Hollywood had made him wealthy beyond his wildest
dreams, McQueen's impoverished reform school days were never
far behind him. He was sometimes known for being tightfisted
when it came to money.

"Steve was frugal. He would take me shopping for little cotton
dresses at those cheap Kmart-type stores. According to Steve, as
long as they were clean and pressed and you looked good, clothes
didn't have to be expensive. Of course I would pout all the way
home. But then he felt you always had to have fine shoes, so we'd
stop at an expensive boutique for a few pairs of three-hundred-
dollar Italian shoes."

Besides cotton dresses, T-shirts, jeans, and expensive shoes, the
McQueen household was well stocked when it came to "FREE
SONNY" T-shirts. According to Barbara, McQueen had six or seven
of my "FREE SONNY BARGER" T-shirts lying around the house. He
urged her to wear hers in public.

"He explained that Sonny Barger was an important bike rider,
that he was in jail unjustly, and he should be set free. He'd tell me,
'Wear your shirt!' So I wore it, and got more than a few strange reactions and weird looks."

In the 1971 *Sports Illustrated* article, McQueen complained that
Marlon Brando's film *The Wild One* set motorcycle racing "back
about two hundred years," but privately, he felt differently about
outlaw bikers, some of whom frequently visited the Hangar.

"He told me if I was ever stranded on the side of the road, broke
down or in trouble, and if a Club member came by, accept their
help. Trust them. Nothing bad would come of it. They'd do me right.
He even went on to say that if the politicians ran the country like
the Club ran its chapters, we'd all be much freer. I think he admired

Steve McQueen, owner of more than two hundred motorcycles, with veteran motorcycle racer Red Wolverton. *(Photograph by Barbara McQueen Brunsvold)*

their honor system and the fact that with bike riders, a handshake was a handshake and a deal was a deal."

Cars and motorcycles soon gave way to airplanes. McQueen wanted to become a pilot like his father. Just as he did with motorcycles, McQueen jumped into the airplane world with both feet. Flying became one more ultimate rush, and like his stable of motorcycles, he set out to own a fleet of airplanes. If he liked something, he'd buy it. He spent hours sitting on the john with a telephone and a *Trade-A-Plane* magazine on his lap, building up his collection. His taste in planes also reflected his love for the restored and antique, including his two prized Stearman PT-17 biplanes, not to mention his rare 1931 Pitcairn PA-8 mail plane.

McQueen returned to mainstream movie-making in 1978 with *Tom Horn*, a Western filmed in Tucson. Although he was a skilled horseman, McQueen hated horses. As he had told *TV Guide* years before, "When a horse learns to buy a martini, I'll learn to like horses." McQueen brought along his dog, a despised canine that bit nearly every member of the crew. (Eventually the dog would mysteriously disappear.) Formerly notorious for womanizing on the set, McQueen was faithful to Barbara, and the two were seldom apart. The couple camped in a motor home parked over the hill from the film's Western set rather than stay in a luxury hotel. Each night, under Arizona skies, the couple would ride motorcycles, drink beer, shoot guns, and play with the coyotes.

The next year McQueen filmed his final movie, *The Hunter*, released in 1980, which was based on the life of a modern-day bounty hunter, Ralph "Papa" Thorson. It was during the filming of *The Hunter* that McQueen's health began to falter. After *The Hunter* wrapped, McQueen fell ill back home in Santa Paula. His next movie, to be filmed in Taipei, was immediately scratched.

By early 1980, Steve and Barbara were married. Moving out of the Hangar and into a refurbished house in Santa Paula, they held a quiet family ceremony at home. For weeks the tabloids had stalked the couple. To combat unwanted publicity, McQueen posted an employee at the front gate, armed with a shotgun.

"He had night sweats and couldn't sleep," Barbara said. "Something was bugging him. Finally he went to a doctor in Santa Paula, who freaked out and sent him to Beverly Hills. That's when they did the lung tests and found out Steve had cancer."

McQueen was diagnosed with mesothelioma, a fatal strain of lung cancer usually associated with asbestos exposure. While he *was* exposed to asbestos in the Marine Corps, he was also exposed to asbestos through the brake linings of racing cars as well as the insides of his racing helmets. Nevertheless, in Barbara's estimation, the cause of his demise from mesothelioma remains inconclusive.

What was certain was that his days were numbered. After several weeks of experimental cancer treatments in Mexico, McQueen returned briefly to Santa Paula before his kidneys began to shut down. On November 7, 1980, Terrence Steven McQueen died in Juarez. He was fifty years old.

Being a cancer survivor myself, I can say that Steve McQueen was truly a man's man, and one of the very first actors to take motorcycle riding seriously and legitimize it. One of

Steve packing his beautiful wife, Barbara, and wearing his favorite flannel shirt.
(Photograph courtesy of Barbara McQueen Brunsvold)

his most famous movie moments was the famous motorcycle jump in the 1963 movie *The Great Escape*. And while stunt rider Bud Ekins performed the jump on film, McQueen actually duplicated the jump off camera, just to prove he could pull it off.

After his death, traces of Steve McQueen's wealth and personal effects were scattered. Over two hundred and ten motorcycles, fifty-five cars, and five airplanes were auctioned off. In 2001, McQueen's Pitcairn PA-8 fetched $294,000 at an auction outside of Chicago. The same rat bike that terrorized Barbara on the 405 interstate is now on display at a gambling casino in Laughlin, Nevada.

"I burst into tears when I saw it," Barbara said. "It was just like he left it in the garage. His sleeping bag was still on it."

Steve McQueen left behind another legacy: the idea that the highest-paid movie star in the world could still be a regular guy. He wore the same plaid flannel shirt he got from one of his movies over and over until it nearly fell off his back. His hair and beard grew as fast and wild as weeds. His favorite thing in all the world was to hang out with the boys, chug beer, and check out bikes.

"Living and riding with Steve was an E ticket all the way," concluded Barbara, summing up her four wild years with Steve McQueen. "Plus, I loved the person, not the actor."

R.I.P. MCQ3188.

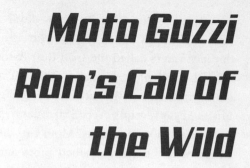

Moto Guzzi Ron's Call of the Wild

Because of my affiliation with the Club, I'm banned from traveling in Canada. The only time I get to see my Canadian brothers is on World Runs. Personally, I have no beef against the Canadian people. It's their cops I don't like. Apparently they don't like me, either. But like the USA, coast to coast Canada has it all: big cities, sprawling prairies, lots of green hills, islands and harbors, plus terrain that creeps all the way to the polar icecap. It's a region inhabited by the tough and the mighty.

Moto Guzzi Ron is an avid motorcyclist from Portland, Oregon, who likes to cover a lot of territory when he rides.

A computer software engineer by day, Ron is not your average Portland RUB (Rich Urban Biker). Ron is a member of a group of endurance riders called the Iron Butt Association. To be "certified" by the IBA for their Saddlesore 1000 run, you need two witnesses (one at the start and another at the end) to sign off on your run, plus documented gas receipts every 350 miles to prove you rode the thousand miles in twenty-four hours. Moto Guzzi Ron falls into the lone- rider category. For a self-described moto-anarchist at heart, going solo beats having to ride only as fast as the slowest guy in the pack.

Ron packs his old lady occasionally, but she doesn't dig riding more than six or seven hours at a time. So she gives Ron his space, as any good woman will do. Off he'll go, for days at a time, racking up fourteen to sixteen hours a day on the open highway.

Moto Guzzi Ron has been riding motorcycles since he was twelve. Obviously, he got his name from his preference for custom Italian racing cycles, specifically Moto Guzzis, built in the same Italian village for over eighty years. Ron tried Brit bikes—Triumphs and Nortons—before moving on to Harleys. He got bored with Harleys, and Jap bikes seemed too perfectly engineered and lacked grit, torque, and personality. He didn't meet the nicest people on a Honda. BMWs didn't have enough edge, either. One test ride and the Italian bug bit him; he's ridden Guzzis ever since.

For cruising, MG Ron rides a Guzzi California EV, which (he claims) blows the doors off most sport bikes, especially after adding new crossover pipes and jacking the bike's horsepower from seventy-four to ninety-one. For long distances, Ron's choice Guzzi is a sleek, bat-out-of-hell black Quota 1100. It's a combination racer/cruiser perfect for open-throttle high speeds and manhandling rugged, uphill terrain and the heavy off-road surfaces you might encounter on a wilderness run. The company tested a nearly stock Quota 1100 and raced it across the blistering Sahara Desert from west to east on a 14,000-kilometer stretch. The bike had the necessary chops for such a long, tough haul and passed the road test with flying colors.

Moto Guzzi Ron's hell-raising motorcycle tale involves a summer 2000 trip he took to the North Pole. His idea was to race his Guzzi Quota into the fabled Yukon Territory and the Arctic Klondike under the Northern Lights and visit towns like Whitehorse, Tuktoyaktuk, and Dawson. He knew the roads were going to be impossible, since not a lot of motorcycles traveled to within spitting distance of the North Pole.

It was late June and the plan was for Ron to haul ass from Portland to Anchorage, then head east toward the Canadian border by way of Tok Junction. He would cross into the RCMP-occupied Yukon toward Dawson. After that, Ron would navigate his way north all the way up to an outpost in the Northwest Territories called Inuvik en route to the North Pole.

We're talking one long summer ride. Ron packed his gear and rode out from under the gray, overcast skies of Oregon toward the Yukon Territory, where the sun never sets in the summer. Portland, Seattle, and Vancouver were a hop, skip, and a jump by bike. Ron blazed through British Columbia and Alaska without breaking a sweat. Soon enough he hit pioneer turf. For a lone motorcyclist, the Yukon can be incredibly desolate. Ron rode for hours without encountering a single person. On the Robert Campbell Highway, he rode 450 miles before he ever saw one other person. He revved up his Quota and found its sweet spot at 90 mph. This was it, no turning back. He was determined to ride up into the Northern Territories and go as far as he could, North Pole or bust.

When you're ridin' high on the open highway, senses heighten as you absorb the sights. Initially, there's an internal dialogue; you talk to yourself. After a while, everything settles down to a cerebral level; you become still. With a 360-degree panoramic view, everything seeps in and registers. The little voice that chatters and natters in your head eventually disappears. Many riders slip into a free form of meditation, except you're much more alert. Far from home, you've got to be ready to be put to the test at any time.

On Highway 8 in the Northern Yukon, the test comes in the form of one of the most treacherous strips of roadway in North America. At first sight, the Dempster Highway is a peaceful, flat stretch of country road. But when Ron pulled up to the entrance of the Dempster, a big sign ominously greeted him: WARNING. NO EMERGENCY SERVICES AVAILABLE. DRIVE WITH CARE. Man, they weren't kidding. The Dempster is 456 miles (734 kilometers) of straight road, basically a huge mound of crushed shale all the way. Unfortunately, shale is a rock formation with razor-sharp edges, so if you ride slowly over the shale, as the rocks and gravel grind under your tires, it gouges out tiny chunks of rubber. Your tires last about a couple hundred miles, tops.

The Dempster, approaching Inuvik, continues on fairly flat terrain, flanked to the west by the Mackenzie River delta. To the east is a stretch of deep green woods dotted with sky-blue ponds, the largest body of water being the Campbell Lake, just south of Inuvik. The road itself is built on a solid bed of permafrost (frozen topsoil). The flattened shale sits ten feet thick on the hardened soil. Parts of the Dempster are rather monotonous

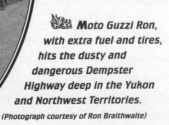

**Moto Guzzi Ron,
with extra fuel and tires,
hits the dusty and
dangerous Dempster
Highway deep in the Yukon
and Northwest Territories.**
(Photograph courtesy of Ron Braithwaite)

to maneuver, and the roadsides often angle off steeply, barely wide enough for two vehicles.

Ron decided that the best way to conquer the Dempster on motorcycle was to do eighty-five directly over the top of the gravel, like a speedboat hydroplaning over water. Along the first hundred miles of the Dempster, Ron ran into a group of guys riding BMW automobiles. They had already turned around. The road was too much for them. Another 150 miles in, Ron met a stranded pickup truck camper who was on his third flat tire. This guy was shit out of luck; he was going to be stuck for days.

The next two hundred miles in, another camper had pulled off the highway. A man and his daughter were crying by the side of the road. A tow truck had shown up, but the driver wanted to charge them $1,500 just to drop the chains, plus an additional twenty bucks per kilometer to haul them out. Since the Dempster Highway is an Arctic refuge area, you cannot abandon your vehicle. The Canadian government operates roving helicopters serving as tow trucks. If they catch you evacuating your wheels, they'll impound the vehicle, fly off with it, and charge you a small fortune to get it back. Canadian highway robbery. Ron had about six hundred bones in his pocket, and he'd be damned if he'd suffer the humiliation of ending his Yukon run in the front seat of some asshole's tow truck. All the scary shit he encountered on the road made him want to dig in and ride faster and harder.

Moto Guzzi Ron navigated the Dempster in one day. At the end of the line, just outside of Inuvik, the locals raised their beer glasses to him. He was the first motorcyclist they had seen all year. The countryside was as amazing as Ron had imagined. The weather wasn't bad—eighty degrees, no rain, and no snow. Off to his left he saw a thundering herd of 25,000 caribou; to his right, bears and moose wandered freely by the side of the road. Above him, eagles flew the open skies with ten-foot wingspans. It was a truly spectacular solo motorcycle run.

When Ron made camp in Tuktoyaktuk, he encountered some of the natives who built creepy little rock piles along the wooded paths. It was something straight out of *The Blair Witch Project*. Plus, the sun never set in Tuktoyaktuk. Twenty-four hours of daylight gave off an eerie feeling of timelessness.

Beyond Tuktoyaktuk, in the farthest reaches of the Northwest Territories, Ron stashed his bike, hitchhiked over the polar region, and took a ride over the icecap in a single-engine plane. Looking down over the miles and miles of frozen ocean, he saw how some inhabitants had created a makeshift highway across the frigid icecap. As tempting and peaceful as the frozen tundra looked from an aerial view, it was no place for a two-wheeler to be sliding around.

But at least Ron could tell his family, man, he'd made it to the North Pole!

As Ron made his way back down the Dempster Highway driving south, a big logging truck approached from the horizon. It had been hours since he had seen any sign of humanity. As the truck thundered past him and blew its air horn (creating an echo through the frighteningly still terrain), the eighteen-wheeler also created an enormous, swirling dust cloud. Ron pulled in his clutch and slowed down as the truck zoomed by him.

The air was perfectly still, and the dust wouldn't settle. The more Ron slowed down, the more he realized he couldn't see a damned thing. In a Zen-like moment of "on the road" intuition, Ron felt the texture of the gravel change beneath his tires. Instinctively, he came to a full stop and looked out into the haze. Another few minutes passed and when the dust settled, Ron looked down and realized he was three feet shy of a two-hundred-foot cliff. Apparently the road had curved off to the left, and he'd taken too wide a turn. Moto Guzzi Ron would have been a goner had he kept riding. Had he sailed over the edge, it would have been "*Arrivederci,* baby," and he and the Guzzi would have been gone without a trace.

By night, the grizzlies roamed the Canadian wild for food. To keep the bears from invading camp, Ron lived on his freeze-dried provisions with no scent to attract hungry bears. Camping along the Yukon River, the next morning he found a giant pile of bear shit not five feet from his tent.

Ron pulled into Fort Nelson, located in the upper, most northern part of British Columbia. Although it was still a rural area, at least they had a Harley dealership in town. Ron felt like he was back in civilization. It was 11:30 at night, the Fourth of July, when he gassed up at a self-serve in Fort Nelson. Alert and ready to roll, another biker on a Harley pulled up next to him. Ron admired the guy's ride. He in turn had never seen a Moto Guzzi Quota. The two got to talking and the guy asked Ron where he was planning on crashing for the night. Ron told him he would take advantage of the twenty-four-hour daylight, ride a few more hours, and then camp out just off the road. The Harley guy turned deadly serious.

"You know this is dangerous bear country."

"I've been out in bear country for the last two weeks," Ron answered.

"No, man," the guy said. "I'm not fucking kidding. You'll never believe what happened to me and my buddies just a couple of days ago."

And the guy told Ron *his* hell-raising motorcycle story.

The guy and his four buddies had been biking through the Yukon with their girlfriends riding on the back. They were headed down the Alaska Highway, forty-five miles outside of Fort Nelson. Out of the bush a huge grizzly bear hunting for game snagged himself a deer. The deer, making one last-ditch attempt to get away, leaped thirty feet over the brush. In a vain attempt to escape, the bloody deer jumped out onto the highway, knocking one of their women off the back of the Harley. Her man must have been a pretty good rider to keep the bike from falling over after the collision with the deer.

But the woman was knocked out cold. Both the woman and the bloody deer were stretched out on the roadway unconscious.

The grizzly came lumbering out of the brush. He stood four meters tall (fourteen feet high) on his hind legs, with eight-inch claws. He let out one heck of a mighty roar. It was fair warning to the world: this was his kill. Then the bear dropped down on all fours, grabbed both the deer and the woman, and dragged them back toward the woods for an early dinner.

These bikers weren't packing heat. Even so, a .44 Magnum pistol would have only annoyed this motherfucker. So the five bikers thought quickly. Surrounding the bear in a semicircle, they revved the engines of their straight-pipe Harleys, making as much noise as they could until the bear finally dropped both the woman and the deer and ran back into the brush.

Thereby proving once again: Loud Pipes Save Lives.

Ron saw the twisted look of fear and anguish in the guy's face as he told his story. He figured the Harley guy was either pathological or he was telling the truth.

When Ron hit Seattle, he called his old lady. Man, did he have stories to tell. As he rode through Washington, a summer rainstorm hit. Ron slowed down and pulled into a truck stop. Since he was a biker and not a trucker, the asshole who ran the stop refused to rent him a room for the night. So Ron roared back onto the freeway. He fought off the sheets of rain by riding a few feet behind a tractor-trailer rig, following the red taillights. At the Oregon border, Ron kept right on cruising until he hit Portland.

Ron swerved back into the neighborhood, pipes blasting loud

enough to wake up the entire neighborhood. After conquering the Great Northwest, the Yukon territories, and the North Pole and coming back in one piece, Ron felt like the king of the bike riders. After exploring the outermost regions of the Klondike under the Northern Lights, it felt great to be home. He had gotten "the call of the wild" out of his system. Johnny Cash was right; sometimes the best part of the journey is the last mile home. And Moto Guzzi Ron was home, at least until the next call . . .

Bay Bridge
Discount

Once upon a time some Club
members and I were over in Frisco
having a ball in this bar when some
weasel smart-mouthed our buddy
Teddy the Wanderer. So Ted socked him. Apparently this guy had a lot of friends, because
when Teddy smacked him, the whole bar
jumped up.

All right! The shit was on.

It was a pretty good fight. There were fifteen or so of us and about twenty-five of them.
We were knocking people left and right out the
door, Western-style. They would run up and
down the street and gather reinforcements,
which only made the fight better. Fresh victims. Once the locals realized they were all
going to get a little bit more than they bar-

gained for, Cincinnati spotted the bartender on the phone, so he
jerked the phone out of the wall.

"They've called the law, boys. Let's go!"

We hit the sidewalk and started jumping on our starter pedals.
We must have looked like a row of pistons, jerking up and down, up
and down. Someone pointed over to my right, and I looked over at
Johnny Satan trying to push away from the curb. He's got a "trim"
little Frisco virgin on his p-pad that probably tops out at about 230.
Johnny was drunk.

One of our guys bailed off his scoot, grabbed a cue stick from
somebody on the sidewalk, stepped up, and lightly tapped Johnny's
girl on the ear, just to get her attention, you understand. Well,
Johnny must have just been letting out the clutch at that moment,
because when his soiled little dove got tapped, she stood straight up
in the fucking air on his rear pegs.

Back in the day, there weren't many tires made for Harleys.
There was Avon, the stock Goodyear Double Eagle (absolutely
worthless), and the very best available, Beck K-555. Johnny was
running Becks, because an Eagle would have just spun and kicked
to one side. He could have just waited until it kicked the right way,
and been gone. But no, Johnny had traction, way too much traction. When he dropped the clutch, his front end soared skyward.

Johnny's bike shot straight across the street, slamming into a
couple of parked cars. Meanwhile, his little sweetheart had done a
perfect back flip and landed flat on her face on the sidewalk. Had
we been scoring, all the boys would have held up perfect 10's.

We blasted out of there thinking everyone was ready, on board,
and in line. As we were jumping up onto 101, headed for the Bay
Bridge to get back into Oakland, we saw nothing but red and blue
lights wailing in the opposite direction toward San Francisco. We
made it across the bridge, but lo and behold, the CHP were waiting
for us at the tollgates on the Oakland side as well. (This was back in
the day when you paid bridge fare going both ways.)

Pay your toll, motherfucker, and pull your ass over. They had us.

But after the CHP pulled us all over, they really didn't know what the fuck to do with us. We were all sitting on the side of the road when we heard a couple of bikes roaring in the distance. Realizing that Big Dexter and Pee Wee weren't with us, it was them we heard first, and then finally saw. All the barricades were down at the toll-gates, but Dex and Pee Wee weren't slowing down. Each picked a gate and—wham!—little pieces of barricade wood went flying all over the place. The cops never even bothered to try and catch the boys.

Funny. We had always thought that those funky barricades were made of pure steel, but as it turns out, they ain't. They're made of real lightweight cheap wood, and when you hit them just right, running about a six-over front end, your bottom triple clamp hits them dead center. Thanks to Dex and Pee Wee, many a night followed when we chose to blast those fucking barricades to smithereens rather than pay their goddamned tolls.

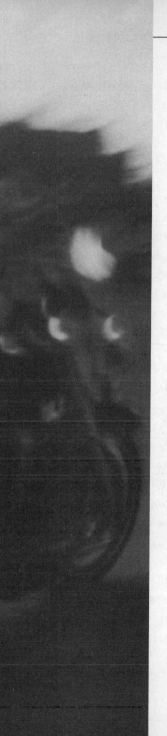

Nan and the Basket Case

Riding motorcycles in New Hampshire and Vermont can be a Jekyll-and-Hyde experience. Of the two states, New Hampshire is friendlier toward bikers. New Hampshire hosts the big annual Laconia bike rally and has no helmet law. In Vermont, on the other hand, cops will write you a $117 tag for no helmet in a New York minute—even if you're riding on the bridge over the Connecticut River that separates New Hampshire from Vermont.

When Hippie Nan first got her New Hampshire motorcycle license, she swore she would never ride until she could afford a Harley. Her obsession started in childhood. When Nan was six years old, her mother dated a smooth bike rider. Mom's Romeo wore an old captain's hat and rode a Panhead FL with huge leather saddlebags that sported dangling fringe and silver buckles. Little Nan went crazy over the rumble and the power and as she grew older, her love of Harleys spilled over into her taste for boys.

"I was the kid who got straight A's in school, but I always went out with the worst boys you could find."

Her first Harley was a coal-black 1986 Sportster 883. It was a miracle the bank lent her the dough. The loan officer shot her a dirty look. Why not something sensible, like a station wagon? The first day she brought the bike home, she could barely ride it off the showroom floor. Her first few miles were shaky, but by the two-thousand-mile mark, Nan felt her confidence set in. That first summer, it rained twelve weekends out of thirteen, but Nan refused to let the rain soak into her road time. In a matter of months, she would put more than 5,500 miles on the boards, cruising the White Mountains of New Hampshire.

Not long after Nan began riding full-time, she split up with her old man. Nan took off to California for a while, and the deal was that her ex would hold on to the bike and keep up with the payments. But when she returned to New Hampshire a few months later, the bank was on her tail for twelve hundred bucks in payments and late fees. She found a quick gig working nights in a factory machine shop, running an automated drill press and lathe. Nan lived on bologna and English muffins. Pretty soon she paid off the bank and scraped up an extra six bills for a rhino chasin' Triumph rigid until she could get back on her feet and find her next ideal ride.

Then one day it happened. Nan crossed the river into Vermont and saw a 1960 Panhead sitting in the front window of a bike mechanic's shop. It was an old bike rebuilt from many spare parts and supposedly in "running condition." The frame was caked in primer. The electrical system was hopeless. The front end was shit. But there was something rustic and gutsy about the '60

A reflective Nan today. Note tattoos and poise.
(Photograph courtesy of Marjorie Goodness)

Panhead that made her want to throw caution to the wind and own a ride like that.

Nan was a good-hearted woman all right, but her tendency for hooking up with older troublemakers was about to cross over into her taste for bikes. Her fascination with the Panhead was a lot like how Nan muddled through life in general. Time to jump in head-first into another risky relationship. She could have groveled for another loan and picked up a newer bike. But her attraction to this classic Panhead was a simple case of mechanics (art) imitating life: Nan (the good) was falling for the bad and the ugly . . . and the un-reliable.

"How much do you want for this bike?"

The guy at the shop was a bit of a whiner; he hemmed and hawed. Well, it wasn't exactly his to sell. (What's it doing in the

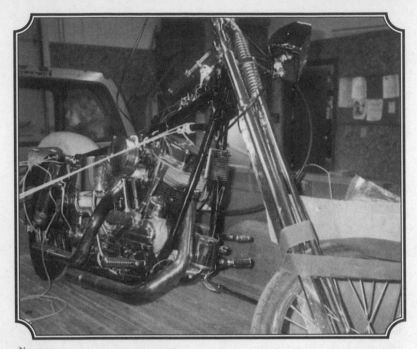

Nan's old '60 Panhead the day she brought it home.
(Photograph courtesy of Marjorie Goodness)

window?) The guy it belonged to had a lot of money tied up in it. (How much is a lot?) A new transmission was on order. (When do you expect it to come in?)

"Find out what he wants for it."

After much bouncing around between owner and middleman, a price was finally fixed on the Panhead—$3,500, and not a thin dime less. By this time, Nan was riding the Panhead of her dreams in her sweet little head. She had to have it.

"Ask him if he'll take my Triumph in trade, plus cash."

The bike's owner danced around the sale for another week or so. Finally Nan couldn't take it any longer. What did she have to do to make this thing happen?

"Hell," she told herself, "I would have fucked Howdy Doody for that bike." So Nan lured the guy into a sleazy motel and gave up the booty.

Nan delivered the cash and the Triumph and loaded the Panhead on a truck (ignoring the sonofabitch's knowing wink), having convinced herself that she'd just scored the deal of the century. But after tearing the bike down and reassembling it at a friend's shop, she would soon face the terrible truth. Kick, kick, kick. Nothing happened. Kick, kick, kick. Nada, zilch. Oh, that sinking feeling. Her '60 Pan officially became a basket case again.

Somebody suggested she take the bike a few towns over to a miracle man named Mook the Healer. Mook had a bike shop ninety miles south of Nan's place. Walking into the shop, she found a textbook biker grease monkey. Blowing his nose into a black railroad worker's hanky, Mook took one look at the Panhead, another at a bum-kicked Nan, and launched into a spitting tirade.

"Oh no, not another Panhead. Fucking junk."

Mook the Healer had quite a history. He'd done a few revolving stints in the county jail and regularly fucked people up when he had a few too many. After leaving the outlaw club he rode with for years, he got married and moved down from Connecticut. He settled down and made a name for himself doing quality mechanic work. He did okay, plus he wasn't drinking anymore. Yes, he was crude, but when he said he could fix something, he damned well would. Mook could cast parts you couldn't imagine, much less buy.

After his spew of obscenities, which included a few jailhouse phrases Nan hadn't heard before, she finally interjected, "You know anything about Panheads?"

"Lady, I hate those fucking bikes."

"I'm not asking you to marry it, can you make it run?"

Mook's face softened. He knew that she knew that he knew she'd been ripped off. Maybe it was the desperation in Nan's face. Maybe he took pity on a basket case. Maybe it was a Friday. Whatever, Mook knew a near-impossible challenge when he saw one and this time he took the hook.

"Tell you what. Leave the motor, and I'll see what we can do."

Emphasis on "we."

The very next weekend, Nan made the pilgrimage back to Mook's shop. Together they tore the Panhead down. What they found was an absolute cluster fuck. No wonder the goddamn thing wouldn't run. The motor wasn't all there. The connecting rods were in backward. The idler gear in the lower end had a left-handed idler screw that had broken off and was stuck back in with putty. (Mook practically hurled at the sight of that!) There was no way this bike was going to run properly. It was basically shit, a basket case indeed. The only good news was that the trannie was good to go. Hell, it'd better hum; it was brand-fucking-new. Mook shook his head in disgust.

"Lady, you got fucked."

"If you only knew" was Nan's comeback.

Mook still wasn't going to let Nan off easy. Getting this thing up and running meant *she* was going to feel every ache and pain. Most nights, he would work on the '60 after he closed up shop. Mook insisted she be there as they sorted through the bike from stem to stern. He showed her what was broken, what was fucked up and why, and what could possibly be salvaged. Then he sent her on the scavenger hunt from hell for parts. There wasn't anything "after-market" that was going to save this basket case. Nan would scour every bike shop from Antrim to Wolfeboro, and practically every rat-trap whistle stop in between.

But Nan didn't blink. She got tough and learned from the master, and did all the legwork necessary, even when it meant taking a mountain of shit from her friends for all the time and money she'd already spent/wasted on the Panhead.

Nan was on a mission to save the Pan. Whatever it took, she'd do. Whatever part she needed, she'd find. Whatever screw had to be screwed, she'd screw. When Nan finally located the last gasket for the basket, her moment of truth had finally arrived. Even Mook sounded stoked when Nan got the call.

"Be here tomorrow morning at eight o'clock. Sharp."

She called in sick at the factory, hopped a Greyhound, and hit Mook's garage that morning with minutes to spare. Mook had wrenched the bike like a symphony conductor. It was over a bucket of KFC and a few six-packs of Rolling Rock that Mook and Nan put the finishing touches on the Pan, and even Mook had to admit, damn, that thing was looking pretty dapper. The starter mechanism was so in sync, you could grab the kickstarter in your hand and kick it over. The black bike with silver trim ran as good as it looked, so the two of them decided to snap a picture of the Pan and send it to a magazine—*Outlaw Biker* was Mook's suggestion—to see if they would actually print it:

Dear *Outlaw Biker*. I'm sending you these pictures because I bought this bike as a "Running Basket." Unfortunately the only thing that ran was the pig fucker who sold it to me. P.S. I'm writing you on this KFC bag because I've spent all my money on this bike and I can't afford stationery. Sincerely, Hippie Nan.

Nan rode the sonofabitch home, stoked out of her mind, tires barely touching asphalt. Although she was wasted from all the late nights she and Mook had pulled at the shop, working on the Panhead proved a valuable experience. She learned a lot in the garage with Mook, getting dirty, wrenching up, drinking beer, figuring out the whys and the wherefores of the bike's psyche. Truth be told, Nan was a little let down when the work sessions ended, but the

Move over, Michelangelo. Nan's Panhead after the final restoration.
(Photograph courtesy of Marjorie Goodness)

Panhead achieved immortality when *Outlaw Biker* ran its photo, one of the first photos of a woman's bike they ever printed.

As time passed, Nan went back to a new Harley. She snapped up a '99 Low Rider from a girlfriend who dropped it on her first ride out, scaring herself to death. But the Low Rider proved to be just too much bike for Nan as well.

"When I look down and I'm doing ninety, I'm going too damned fast. I knew if I didn't get rid of the Low Rider, I'd end up dead. Plus I couldn't even change the oil myself."

Soon it was back to rebuilding bikes when Nan swapped the

Low Rider for a classic 1970 FLH Shovelhead Police Special. She traded in the sidecar for a motor rebuild. Finally, another bike to work on, but a much more dependable ride than a Pan. On an East Coast run to Florida, the refurbished Police Special ran like a watch. The ride from Titusville to Daytona at six in the morning, with the sun coming up over the ocean, gave her goose bumps.

Nan continues to tweak the FLH toward perfection. As for the Pan, long may it run, but even Nan must admit, not many chicks can ride and restore old Panheads. They're too freaking undependable. Since Nan does most of her riding alone, she'll stick with the FLH.

Nan learned a lesson from Mook about both bikes and life. Some relationships, like motorcycles, are worth fixing and fighting

Nan's current ride, a reconditioned 1970 FLH Shovelhead Police Special sans sidecar. (Photograph courtesy of Marjorie Goodness)

for, even when, on the impulse, you probably jumped in way over your head. But commitment takes patience and a steady hand. And you know what they say about Harleys: short of melting them down to the ground, a good Harley will always rise from the dead.

Nan recently rode by Mook the Healer's bike shop. It had been a few years since they worked on the Panhead together. The door was locked and she pressed her face onto the front window; the building was dark and empty inside. As Nan started the FLH, she gave Mook a mental high five and rode on her way back home.

Draggin' the Line with the East Bay Dragons MC

In 1959, two years after my club hit the
Oakland streets, another club was born in
East Oakland. Since '59, the East Bay
Dragons MC has been one of the most re-
spected African-American motorcycle clubs in
the land. Throughout my hell-raising motorcy-
cle days in Oakland, I crossed paths with the
Dragons many times.

Time flies and it's still kind of hard to believe that a club like the East Bay Dragons is
fast approaching the fifty-year milestone.
They've functioned consistently under the lead-
ership of just one man, Tobie Gene. Tobie has
been the Dragons' one and only president since
the club's formation. In the beginning, the East

🔥 *The East Bay Dragons MC outside of Helen's Bar-B-Que. Oakland, September 1966.*
LEFT TO RIGHT: *Joe Louis, Tiger Paw, Albert Guyton, David Bird, Corky, Tobie Gene, Jay Pettis,*
Bags (LYING DOWN), *James Hooker, Butch, John Smith, Willie "Poor Hop" Harper, Wally Eps,*
Aubrey Wesley. (Photograph courtesy of the East Bay Dragons MC)

🔥 *Tobie Gene* (LEFT)
with an East Bay
Dragons patch-holder.

(Photograph courtesy of
Paul Butler)

Bay Dragons originally cruised the scene as a car club. But their transition to bikes was a quick one.

"Most of the original guys were married and were raising families," Tobie Gene recalled, "and it wasn't like today where we have two or three automobiles per household. Back then there was only one family car."

That caused problems in the beginning for the founding members who needed to attend club meetings and functions but had old ladies at home who used the car to keep the family functioning. Out of necessity, two or three of the original Dragons started riding motorcycles. Tobie was already making his way around East 14th Street on his 1951 EL Panhead.

"Finally, somebody stood up at a meeting and said, 'Hell, we oughta just go out and get us some motorcycles and leave this car stuff outta here."

And that's when the East Bay Dragons shifted gears and officially became the East Bay Dragons MC, decked out in red, green, and gold patches on the backs of their leather jackets. Before the club found itself a suitable clubhouse, the Dragons met at Helen's, a barbecue house at 85th and G, not far from the original Black Panther Party headquarters. Helen's had a seating area off from the main counter, and that's where the East Bay Dragons MC held its early club meetings. Any time the club was caught between clubhouses, they could always count on Miss Helen to welcome them back with open arms and some of Oakland's best BBQ ribs.

To this day, much of the car club mentality has stuck with the Dragons. While Tobie Gene stressed safety for the new riders, at the same time racing and designing fast bikes were important ingredients in the club's formative years. During the early days, members got around the expense of owning a Harley by building their own bikes. The rebuilding ritual usually took place at Tobie Gene's oversized garage. Everybody was a mechanic and had his own bike in a box. A makeshift assembly line was formed and the chopping process began. One club member, Johnny Mendez, was always in a hurry to put his stuff together, and he generally set the pace as the rest of the Dragons worked fast and furious to catch up. The East Bay Dragons are still a competitive bunch when it comes to the physical appearance and performance of their bikes.

"We tend to have a competition going," said Filthy Phil Baker. "Who has the best paint, the best chrome, the fastest runner, the best sounds, or the most miles? That's how we egg on our younger riders. Our club is about paint and chrome. You've got to keep your bike up and running, and sometimes that's an expensive proposition. We spend lots of money on quality chrome, special paint, and things like whitewall tires."

Unlike most African-American bike clubs, from the get-go the

Filthy Phil Baker leaning on his ride. (Photograph courtesy of Paul Butler)

Dragons trashed the notion of putt-ing around East Bay streets on full dressers. In the early days, you'd *never* see an East Bay Dragon riding a full dresser with a raccoon tail tied to a radio antenna. Instead, members chopped their own bikes, tooled their extensions, and built their own front ends, often at campus machine shops, particularly at downtown Oakland's Laney College.

"We didn't like the full dresser Harleys," Tobie Gene admitted. "We called them 'garbage wagons.' Man, we were choppers all the way."

Like a lot of street riders in the 1960s, the Dragons bought full dressers and stripped them down after a lot of cutting and welding. Saddlebags, which sell today for hundreds of dollars, ended up floating out in the Bay. A twenty-one-inch front wheel with no front brake looked just fine. That is, until the Oakland PD stepped into the act. Then came the tickets for no front brakes. Funky turn signals. Ape hangers and T-bars. License plates mounted sideways. One rearview mirror instead of two. While the club made its peace with the OPD, it wasn't unusual for the Dragons to roll into neighboring towns only to be "welcomed" by the cops with warnings like, "If your feet hit the ground, you're going to jail."

Motorcycle riders were one thing, but a raging pack of thirty or more Dragons on choppers rolling down Main Street sent many a citizen behind closed doors.

"Every time we rode into San Leandro," Tobie recalled, "as soon as we left 105th Avenue and by the time we got to Davis Street, the cops had already stopped us and handed out tickets. They'd tell us to get back to Oakland. And they had more police waiting at each block until we left.

"One time we were crossing the Bay Bridge and I mean we were tooling. The Highway Patrol pulled over the whole pack. All of us had T-bars and front ends standing way up. The patrolman stood there looking at the twenty-ones, wondering how the hell our bikes were standing so tall in the front. It turned out that we added some slugs on the front end, which put a three-inch rake on the gooseneck. We all just stood back and laughed. The cop never did figure out how we got our bikes that tall."

The East Bay Dragons' chopper-riding notoriety flourished. Just one member (an original co-founder) was permitted to ride a full dresser, and he was the man who carried all the tools and conducted the necessary business on runs. Then the club voted to become a Harley-only group. Even today, if you're a Dragon, you can ride a Sportster, a Shovelhead, a Road King, whatever, but if it isn't Harley, don't even think about it.

It's a rule that stands today, but at the time it was voted on, it was sometimes difficult to find decent bikes. During that time (1969), I always kept a line on a few available Harleys, but I usually saved them for my own club members. Some riders, including Tobie Gene, considered throwing in the towel and switching over to another brand of ride.

"We were having a hard time finding affordable Harleys," said Tobie. "That's when I sold mine and decided I was gonna get me a [Honda] 750. Then one day I came home, and eighteen Dragons were sitting in front of my house. I thought somebody had been killed, so I jumped out of my car, ran over, and asked, 'Hey! What's going on, guys?'

"We understand you're going to get yourself a 750."

"I sure am," Tobie answered.

In 1969, a Honda 750 (just introduced) cost $1,300 and Tobie had $2,400 burning a hole in his pocket. He knew he could pay cash for his ride and be done with the whole Harley hassle.

"We had a meeting and I was the rabbit in the brier patch," Tobie remembered. "Everybody was upset, especially Bags, our vice president. The Dragons *hated* Japanese bikes with a passion, especially those little 350s and 450s. Those bikes were for that *other* club across town, the Vagabonds."

Out of respect for Tobie Gene, the club discussed a quick resolution. But the way VP Bags saw things, if Tobie rode a 750, that

🔥 *Bags, a longtime Dragons VP, wrenches his Reno Express.*
(Photograph courtesy of Paul Butler)

could only open the door for the worst possible scenario. Guys who rode 750s undoubtedly had friends who rode 350s and 450s, and what if those guys graduated to 750s and, worse yet, went on to become East Bay Dragons?

And that's when Tobie Gene called for a vote.

"So what's it gonna be? Are we gonna ride Harley-Davidson choppers, or are we going to be a mixed club with bikes like 750s?"

The vote came down for Harleys all the way, that is, except for co-founder Albert Guyton, who had just bought his Honda 750 in Alameda the Tuesday before. Tobie stuck with Harley.

"We made an exception for Albert because he was a good man. But when he retired, Lord, you couldn't even talk about joining this club with a Japanese bike."

I **was seventeen when Tobie Gene lived not too far from me, on** 23rd Avenue. He was twenty years old, and we'd meet over at the Doggie Diner and have coffee with his younger brother, Joe Louis. We got to know each other pretty well. A couple of years later, I rode a customized Harley I called the Orange Crate. It was an Oakland-Orange Knucklehead M74, chromed out and built for speed. At the time Tobie was still riding his EL Panhead. His wasn't a seventy-four; it was more like a sixty-one-cubic-inch motor. But Tobie Gene really dug the Orange Crate. I told him I'd sell it to him for seven hundred bucks.

Damn. That was a lot of dough for those days. Tobie's EL had cost him half that amount. The guy he bought it from was a neighbor named Fletcher. And Fletcher didn't really own enough of the bike to actually sell it. It turned out he hadn't been making his payments, so when Tobie Gene brought the bike into the Harley-Davidson shop on 82nd Avenue to get it fixed, Cliff, the shop's owner, came out yelling, "Get off of it!"

"Get off what?"

"That motorcycle. I've been looking for that bike for almost three months."

Never mind buying the Orange Crate. Tobie was left walking . . . and out three hundred and fifty bucks. After some hard threats, Tobie got his money back from Fletcher and switched over to a Sportster. But the craziness didn't stop there.

One day Tobie Gene and the Sportster were doing about fifty-five. He had just washed the bike, and when he hit the brakes, nothing happened. No grab. In order to stop, Tobie had two choices: hit either the house on the corner or a parked '53 Chevy.

Tobie chose the house.

He tried to lay the bike down, but the Sporty kept on course. Tobie finally let go when he hit a high curb. But the bike kept on going. "Damn," he thought, "no insurance and this two-wheeled mechanical varmint was headed straight for some little old lady's living room."

Miraculously, the bike hit a tree and as Tobie watched in disbelief, the riderless bike continued to orbit the tree as if a ghost was onboard steering it in circles. Then, as if the Sportster had a mind of its own (typical of so many Sportys), the vengeful bike came back around as if to say, "Okay, Tobie, you fool, you're standing there looking at me and doing nothing about it. I'm gonna come back and bite you."

And so it did. Tobie's Sportster got him on the leg, and when his wife came home that night he was still shaking in his bed. When she pulled the covers back, his knee was three times its normal size. Tobie lived in a cast for weeks, with torn ligaments stretching from his ankle to his knee.

Some of Oakland's finest black motorcyclists have ridden with Tobie Gene and the East Bay Dragons MC, including Rev, Jack Green, and Mo Holloway. The roar of their Harleys attracted the attention of many young black men from the East and West Oakland neighborhoods. As schoolkids, guys like Filthy Phil Baker and Byron "Bap" Baptiste watched in amazement as members in leather and denim revved up and down Oakland streets, riding wickedly customized rides. They vowed that one day they, too, would wear the red, green, and gold patch.

"Once I got out of school and went to work," said Phil, "I saved my money and eventually bought a bike. Then I went straight down to Miss Helen's and joined the club. That was in 1970."

Phil was one of the last Dragons to join during "the Miss Helen days." After three months of prospecting (running errands and hanging out), he was in. Eventually Miss Helen became ill and closed her barbecue joint. Then the city tore the old building down. These days, the club maintains a nightclub-sized building on East 14th Street, one of Oakland's main drags.

Willie Harper and Willie Josh were cruising East Oakland in a 1963 Corvair when they first came upon the sight and sound of three Dragons screaming down the Cypress Freeway. Their choppers were chromed to the hilt. They were wearing jeans, shiny black patrolman's boots, and rough-cut denim vests over their leather jackets. Willie stepped on the gas for another look but soon lost the flashy bikers.

Not long after, Willie dropped in on the Dragons' headquarters wearing a leather jacket and riding a Honda 250. He had hoped to impress the club members with his new look and ride. Predictably, he was met with ridicule.

"They all laughed at me and took turns riding my 'Schwinn with a motor in it,' " Harper remembered. "Then they said, 'Come back, son, when you get a Harley.' "

Willie Harper was happy to lose the 250. It had already cost him thirty-two stitches after he dumped it on the highway. Then he got lucky. Willie scored a '57 Panhead for eighty-five bucks! With a little coaching from a friend, he chopped the bike and took the peanut tank over to Tommy the Greek for a custom paint job. Out of money, he sprayed silver paint over the parts he couldn't afford to chrome.

Now he was ready to confront the Dragons once again.

The club continued to test Willie. After their meeting, Joe Louis and Sonny Wash followed him home and tore apart the '57 Pan, spreading the parts all over Willie's garage floor.

"Call us when you get this thing back together," they said, riding off.

Just then Tobie Gene happened by. He understood Willie's dilemma and offered him a hand.

"Bring it by my place, and we'll get this thing back together for you," Tobie offered.

Together they reassembled the Pan piece by piece, screw by screw. Willie not only learned firsthand about motorcycles, he now had a bike fast enough to whip Tobie, Lulu, and Hooker on the open freeway. Thirty-plus years later, Willie, now dubbed "Poor Hop" by the other members, remains a member in good standing.

Before he became an East Bay Dragon, Brother Van frequently took his wife and two kids out on Sunday drives. As they cruised down East 14th, they stopped at a traffic light at 94th Avenue. Three Dragons—Bags, Jay Pettis, and Poor Hop—rumbled to a stop right next to Van's car. Van lived in Oakland, but this was the first time he had seen these characters. So he yelled out of his window to the bikers, "Hey! Can you guys do a wheel stand for my kids?"

"No. We don't do that stuff anymore."

The light was still red.

"Would you do one for my wife?"

"Naw."

Seconds before the light turned green.

"How about doin' one for me?"

As the light turned green, all three bikes flew up in the air and popped wheelies. Ever the showman, Bags even *shifted* his bike in midair.

And that's when the motorcycle bug bit Van. Van's wife immediately read his thoughts and took the offensive.

"You don't need one of those," she said.

But it was too late. His head was in the clouds.

"God," Van thought, "I've got to have one of those."

Van's first bike was a super-raggedy Panhead. It had no outer primary cover, the seat was rickety, and it was rusty as hell. But it ran. After Van began wrenching the Pan, he soon joined up with the Dragons.

To this day, Van swears nobody can outbuild him. He figures he's built some of the fastest bikes in all of Oakland. ("Only Brother Bap's Touch of Gold comes close.") Most of the Dragons agree. Van's

Van the man and one of his "pretty," souped-up Harleys.
(Photograph courtesy of Paul Butler)

earliest bikes were always pretty, but it was only after Tobie Gene and
some other members raced their bikes at the Fremont Drag Strip
(and whipped Van's ass on the track) that his rides had to be fast, too.

Nowadays every bike that comes out of Van's garage usually
comes out sideways, fast . . . and, yes, pretty.

One day Tiger Paw and Van put together a kickin' '72 Lowrider
basket case for a cross-state trip Van had planned. They finished the
bike on a Sunday morning, and Van wasted little time heading out.
He rolled the Lowrider out of Oakland shortly before noon and hit
the highway. An hour or so later, at the Highway 5 junction, Van hit
a little dip and bounced. Off flew the carburetor. The bolts had
worked their way off by vibration. Desperate, Van collected the
carb, walked out into a cow pasture, found some baling wire, and
wired up the carburetor so tight he could practically stand on it.

Back on the road with a vengeance, Van throttled that Lowrider relentlessly, chasing through Texas, Oklahoma, and all the way back without a single breakdown. But it was somewhere around Albuquerque in the dead of night when Van began to grow tired and sleepy. A mixture of exhaustion and wind chill tested Van's stamina. He could see the city lights way off in the distance, and just the thought of reaching the town kept him from dozing off at the handlebars.

Just then a round dark object came into view on the pavement. Whatever it was, Van swerved and narrowly missed it with his front wheel. Then—*bam!*—the object hit Van's foot. He felt a sharp slicing pain as he made contact with the object. Panic set in. At first he thought he might have lost his whole foot. What happened? Van needed to get into town before he bled to death. But he quickly changed his mind and pulled the bike over. He took his boot off and literally saw his foot pulsating—already sore and swollen—except there was no blood. Nearby, the object rolled over to the side of the road.

What did Van hit?

It was a dead armadillo.

Like most bike clubs, the East Bay Dragons MC has seen motorcycles come full circle. As is the current fashion, the touring bikes and full dressers that were scoffed at years ago have since replaced most of the choppers of yesteryear. Riding comforts like windshields have taken away some of the neck and shoulder pain that comes with fighting the winds at 90 mph during a long day's run. Gone are the days when you wrenched your own Harley-Davidson on the roadside. Because of modern amenities like electronic ignition and fuel injection, computerized service centers and authorized dealerships are the gatekeepers for a lot of today's bike maintenance.

As for the Dragons, they've seen their share of loyal members pass

Prez Tobie Gene (RIGHT) **with Byron "Bap" Baptiste.**
(Photograph courtesy of Paul Butler)

on. Most notably, Brother Bags was killed in an accident in 1999 coming back from the annual Street Vibrations run. Other fallen brothers include Shorty, Mighty Quinn, Wild Man, Burlermore, Bobby, Big Mac, Sam, Donald, Ken, Juice, Pete, Steve, and McCoy.

Thirty- and forty-year members like Tobie Gene (who has raised three kids and sent two to college), Joe Louis, James Hooker, JC, Sonny Wash, Ezell, Big Al, and Sweet Lee still make the annual runs to Fresno, Los Angeles, Reno, and Las Vegas. Longtime members like Paul, Eddie, Stranger Man, Cat Man, Charles, Cat Daddy, Lil' Mac, Dirty Red, Pretty Tony, Joe Joe, Super Coop, Leon, and Melvin join the "newer" faces like Dave, Car Wash, Fuller, Boopie, Fat Dragon, Crazy, Limo, Pac-Man, Webb, Lil' Al, Capt'n Hook, and G-Man on the highway. For the past twenty years, members have represented the club at Sturgis and other national runs. It's been said that while many members of the Dragons may have come from various MCs, once they join the East Bay Dragons, no other club will do.

Black bikers have cultivated a strong tradition, as African-American motorcycle runs and events like the annual National Black Riders Round Up now attract upwards of one hundred thousand bike-riding enthusiasts. Scores of bike clubs (racially mixed) like the Soul Brothers, the Chosen Few, and the Defiant Ones continue to attract new black riders.

Yet few clubs can match the longevity of the East Bay Dragons MC. Since 1959, they've remained highly visible on the streets of Oakland, as they close in on half a century of ridin' high and livin' free.

Jerry's Kids

Ever wish you still had your very first Harley? What if you were smart—or dumb—enough to keep all your bikes? Imagine being able to recall every hell-raising moment in your bike-riding life by stepping out into the garage and—ta-da!—there they are, virtually in a line, all of your bikes—your kids—bringing back all the misadventures, the scrapes with the law, and the high times you spent riding. To be honest, you'd either have to be rich or crazy. In Crazy Dave's case, it's the latter.

The 1954 Panhead. Crazy Dave grew up in a strict Seventh-Day Adventist home, a family ranch in the Central Valley of California. Dave's father was a heavy-equipment operator who worked mainly on urban construction sites in San Francisco. With his father always gone, it meant that a lot of the responsibilities around the ranch fell on Dave.

Dave has been Crazy Dave ever since he got his first ride, a Bonanza minibike (minirocket) equipped with a 100 cc dirt bike engine and five-speed transmission. Soon he made a name for himself—Crazy Dave—jumping ramps and fences on the thing.

At age fifteen, it was Dave's job to haul the hay around the family ranch. Dave's father left behind the old, rusty 1963 family Ford pickup for him to do this. One day at the feed store, one of the grizzled locals made young Crazy Dave an offer he couldn't refuse. The local needed a pickup but had little money to buy one. Would Dave consider swapping the family relic for a motorcycle?

Motorcycle. Just the word was bound to perk up the ears of a gawky fifteen-year-old. Dave was no exception, so he followed the old man back to his ranch, and sure enough, there it sat in the barn, a full-dresser Harley-Davidson, a magnificent 1954 Panhead. The terms of the deal were simple: a straight-across trade, the bike for the truck. Like a fool, Crazy Dave pondered the offer.

"Man," Dave told the old man, forecasting his impending doom back home, "my dad is gonna kill me."

The old man thought for a minute and interjected, "Not if I throw in a hundred bucks. Then you'll have made a *real* good deal."

The bike hadn't run in years, and it took almost four hours for Dave and the old man to get it humming. After a few quickie lessons, Dave was riding the bike around the old man's yard like a rodeo clown, dodging barrels, nearly tipping the Pan over several times. But once Dave was able to keep the bike up and straight, the old man followed him back to the ranch. They dug through the family files for the pink on the truck. By Friday, when Dave's dad came home, his first question was, "Hey, where's my truck?"

What Dave didn't know about riding a full-dresser Harley, he knew about the lookin'-cool part. With his *Wild One* leather jacket and newly greased-back hair, Crazy Dave transformed into Marlon

Brando's Johnny. One Sunday morning he rode the '54 over by the house of a neighbor, a dairy farmer more interested in his Sunday paper and none too impressed with Dave's new ride. The bike was parked right in front of the old man's living room picture window.

When it came time to make his exit, Dave learned that there was more to riding than just being a poser. As he kicked down hard on the starter, the motor fired up but the compression stroke went ballistic. The backfiring '54 hurled Dave upward and bucked him right through the plate-glass window. Landing in the farmer's living room, glass strewn all over, Dave picked himself up. The dairy farmer barely peered over his Sunday paper. Clenching his pipe in his teeth, looking over his bifocals, he casually asked, "Dave, I know you're gonna be okay, but are you staying for breakfast?"

Then it was back to his pipe and paper.

The 1956 Panhead.   It didn't take long for Crazy Dave (at eighteen) to fall in with some local Central California outlaws. The cops picked him up for drag racing, and this attracted the attention of Indian, the Swede, and Little Ugly, three members of a club called the Barhoppers. Crazy Dave was voted in as a prospect.

By this time, Crazy Dave was up and rolling on a basket-case '56 Panhead. After totaling it once, Dave took the original engine and transmission and transplanted them into an old Harmon frame with an inch-and-a-half rake and a sixteen-overstock Springer front end. The old Hardtail, boxed-in frame and all, was a mover.

"It wasn't a show bike, but it wasn't no slouch, either. Lots of chrome, too."

Coming out of La Grange and up into the hills by Sonora, Dave, his pal Frank, Indian, the Swede, and Ugly had assembled an ad hoc confederation of fifteen riders. Riding in the pack were three small-town clubs that actually hated one another. But nobody flew colors that day; everybody was riding to be riding. It was a nice day for a mixed crowd. Nobody was out for payback.

Coming down Dead Man's Curve (every town's got one), Crazy

Dave and the herd were putt-ing about eighty miles an hour. Everyone downshifted after some signage warned about an upcoming ninety-degree left turn at the bottom of the grade. Well, everybody downshifted except for Frank. Dead Man's Curve ended at a tight barbed-wire fence cordoning off a cow field. Frank raced downhill on his Sportster and didn't make the turn. His bike ran off the road and straight toward the fence.

"Frank didn't slow down, he just kept going at sixty miles an hour. The barbed-wire fence literally took his head off."

The pack watched in horror as Frank's head went straight up in the air—helmet and all—and bounced back down onto the ground. One guy vomited at the sight. More than a couple of outlaws scooted off quickly, scattering before the cops and ambulances arrived on the scene. The freak accident freaked out the Central Valley citizens for several years until it settled into a rural myth. But if you ask Crazy Dave, he swears it really happened.

"Man, I was there."

The next weekend at the Barhoppers' clubhouse, things got tense. During a particularly nasty argument, one member shot another in the stomach. Little Ugly pulled Crazy Dave aside and gave him some free advice.

"Dave, you don't need this bullshit. You're too young. Stay crazy, but go join the Army or something. Get the hell out of here."

That was on a Saturday. On Monday morning Crazy Dave joined the Air Force and put the '56 Pan on ice until his return.

The 1967 Shovelhead.   After a hitch of flying planes and playing pilot, the first thing Dave did when he got out of the Air Force was to buy a '67 Shovel. Crazy Dave was still crazy. In need of quick cash, he got himself involved in some gunrunning in Florida with an outlaw buddy. Crazy Dave hopped on his Shovel and made the cross-country run to do the deal. Once he arrived, he stashed the Shovel, jumped into a brand-new Corvette, and headed toward Fort Lauderdale to drop off a shipment of Uzis stashed in the 'Vette's trunk. The guns, while (technically) purchased legally in a Mid-Atlantic state, were to be sold off to some shady Jamaican gangsters.

Flying down I-95, Dave had the 'Vette's top down. The stars put on quite a show that night. It was somewhere around Jupiter, sixty miles out of Fort Lauderdale, that three state cops pulled Crazy Dave over with guns drawn. "Shit," thought Dave, as he was told to keep both hands on the steering wheel, "somebody must have ratted me out."

After a quick peek inside the trunk, Crazy Dave's next stop was Dade County Jail. The multitiered jail structure housed some of the baddest felons in America, including a few prominent Colombian drug lords. Not far away, no less than the former Panama dictator Manuel Noriega held court in his cell. But Crazy Dave didn't have the juice (or the notoriety) for solitary confinement. With twenty guys to a cell, Dave made the best of his incarceration by getting to know the drug boys inside. Not bad guys, really. Once it was imminent that Dave would post $200,000 bail, some of them passed on slips of paper with names, phone numbers, and messages for the outside.

This was picked up by the video surveillance cameras inside, but the judge seemed willing to drop all the gunrunning charges if Dave would hand over the slips of paper. Dave said "No way" and a strip search found nothing. But Crazy Dave was small potatoes for the Dade County ATF. They had bigger South American fish to fry. Dave's lawyer pleaded him out with a $2,500 fine. Because the guns were legally purchased (with the receipts to prove it), the court offered to give him back his Uzis. But Crazy Dave was eager to put Florida into his rearview mirrors.

"Man," Dave told his lawyer, "they could melt those fuckin' guns into a statue of Jesus Christ for all I care."

Crazy Dave's gunrunning days were over before they started; he couldn't wait to get back on his Shovel and hightail it west to Cali-

Crazy Dave sittin'—not truckin'—with his beloved 1999 Wide Glide.
(Photograph courtesy of Dave Daugherty)

fornia, *muy pronto*. He felt well out of his league among those Miami *Scarface* motherfuckers.

The 1999 Wide Glide.   These days, Crazy Dave divides his time between riding his 1999 Wide Glide on the Pacific Coast Highway and, yes, impersonating Jerry Garcia. You see, Crazy Dave in his forties bears an uncanny resemblance to the late Grateful Dead guitarist, with the same facial features and curly hair. The only Garcia trait that's missing is the chopped ring finger on his left hand. When Crazy Dave dons a black T-shirt, sweats, and a guitar played slightly out of tune, it's virtually impossible to tell the difference between the aging Crazy Dave and the ultimate psychedelic Deadhead.

The Jerry act started in, of all places, Minnesota. Garcia had just died and Crazy Dave was playing lead guitar in a bar band at the time. Grieving Midwest Deadheads held a vigil for their fallen hero at a place called the Red Caboose Bar. Donning shades and Garcia's trademark black T-shirt, in walked Crazy Dave. The place came unglued. Women humbly came forward, bearing gifts of black roses and teddy bears. When Dave climbed onstage to join the Caboose's Dead cover band for renditions of "Truckin'" and "Casey Jones," the crowd went apeshit.

Pretty soon, Dave cashed in and ramped up the act with dancing skeletons, dry-ice smoke effects, and a faithful version of "Touch of Grey." A few Grateful Dead tribute gigs followed, as well as some modeling stints and TV commercials.

But it's still the Wide Glide that rings Crazy Dave's bell, summing up over thirty years of riding. Staying true to the Grateful Dead motif, the Glide sports a holographic paint job that features skulls and dancing skeletons.

"As you walk by the bike, the Grim Reaper gives you the finger."

Crazy Dave loves his Glide, but the '54 and '56 Panheads and the '67 Shovelhead are "all in the family." Dave swears he'll keep his bikes until the day he dies. They've been there for him through the rough and the smooth. To Crazy Dave, his bikes are like his kids.

Jerry's kids.

And what a long, strange trip it's been.

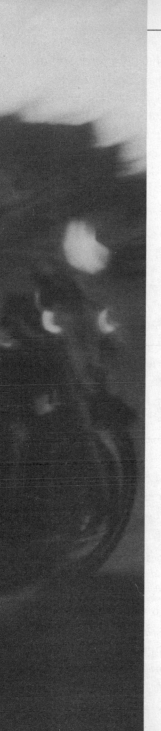

Cros and Cincinnati

A **lot of shit has been written** about David Crosby of Crosby, Stills, and Nash and the Byrds. David is a good friend of Cincinnati's, a friendship that both hold in deepest regard. Crosby has received what one might call bad publicity from a lot of people who should know better. Other than his frolic through the freebase thing (which Cincinnati observed first-hand and did not like for a minute, during which time Crosby kept far away because he knew the score), Crosby rarely makes the same mistake twice.

Of all the rock stars, movie stars, and celebrities we have known, David Crosby is one of the few who is willing and able to cover the ground he stands on by himself. Crosby was

one of the only ones who spoke up for the Club at the Altamont fiasco. He has a wonderful, talented wife, Jan, and another son he's rediscovered who is also a fine musician. We haven't seen him in a while, but he is still a friend and a fine man.

Every year in northern California they have what is called the Redwood Run, put on by the Northern California Harley-Davidson Dealers Association. We had been going there for a number of years, and Cincinnati and Preacher always had an eye on a place up north called the Benbow Inn. The place was bad as fuck, resting on the banks of the Russian River. It was just too cool, so cool, in fact, that they've never been able to get a room there.

Cincinnati was kicking back at the run site when up walks David Crosby. He had been looking for Cincinnati to give him—ta-da!—a room key to a fine room right next door to his own at the Benbow Inn. Cincinnati, pleased to see his hombre and jacked about finally crashing at the Benbow, stuffed the key in his back pocket and then sort of forgot about it.

That night, Saturday, Cincinnati anticipated meeting a group of fine ladies at the top of the hill, just outside the run site, at 10:00 P.M. sharp. These girls all had bikes that their old men, respectable rider types, had bought for them. Each gal wanted a little of the biker fun they had only heard about for years but had never seen or felt. Each year, on the Saturday night, the girls would make sure that they got their old men good and liquored, so that when they passed out, they would roll them into their beds and then take off for the top of the hill to meet their Cincinnati. This little ritual had grown a little out of control over the years into a spicy tradition, swelling to twenty-plus ladies, most of them riding Sportsters.

Cincinnati called them his Sportster Mob. They would usually run up and down the highway together and hit a few bars until 2:00 A.M. Then Cincinnati would show them a few places not on the tourist maps and get them all back to their guys by daylight, usually in one piece.

This particular night, a couple of the girls told Cincinnati that last year somebody had ratted on them about being in the bars without their respective old men. That's when it hit Cincinnati, and he pulled out the key that Brother David had laid on him.

"Hmmm. Benbow Inn. Shiiiiit! Follow me, ladies."

Twenty-something Sportsters (with Cincinnati in tow) rolled up to the inn in search of the lobby bar. Fact was, there was only one critter there, a bored-to-death bartender playing with his nuts. (The ones in the dish.) A million motorcycles parked out front and nobody in the bar? Un-fucking-believable! What's up with that? Cincinnati guessed these pukes with the socially acceptable biker attitudes just didn't like to have a bit of fun.

Cincinnati eyeballed the jukebox and couldn't believe it. Vaughn Monroe. Andy Williams. Johnny Mathis. Not a thing you'd let your dog listen to. He asked the bartender where room so-and-so was, and got the proper directions. Taking a fifty-fifty shot, Cincinnati rapped on Crosby's door, and lucky night, Jan opened the door.

"Jan, darlin', is Cros about?"

"He's down the hall in a friend's room."

Cincinnati told Jan his plight. "The bar has a fucked jukebox, but a piano sits in the center. Any way in hell we could get David down there to tickle the ivory?"

"Hell, yes," Jan guaranteed, without missing a beat, and the two headed down the hall to grab Crosby out of the room.

By now the bar was full of socially acceptable motherfuckers trying to make a move on Cincinnati's Sportster Mob. In a couple hours the place resembled a swinging singles bar. As Crosby took to the keyboard and sang, the RUBs got the not-so-subtle message that this was a private party and it was time to scoot while Cincinnati and the rest of the patrons had a blast until closing. The bewildered old bartender behind the bar didn't know what hit him that night; he resembled a figure from a carnival shooting gallery, moving from one end of the bar to the other, trying to keep up with the drink orders. After last call and closing time, Cincinnati and the Sportster Mob freaked and frolicked till daylight while Cros and Jan went back to their room.

Every Sunday morning, at every Redwood Run, everybody knew where to find Cincinnati. He'd be hanging at his Daly City brother's maggot wagon (catering truck). That year was no exception; at 7:00 A.M., Preacher pulled up to Jingo's maggot wagon, riding all the way up from a function in Berdoo. He was packing his lady, who had been feeling a little under the weather. Preacher spotted Cincinnati. They needed a place to crash.

"No problem," says Cincinnati. "Follow me."

As they came to the turnoff toward the Benbow, Cincinnati crowded Preacher to take the exit.

"There?" Preacher pointed incredulously.

Cincinnati shook his head yeah as Preacher started laughing like a motherfucker.

As the three strutted through the lobby, Jan waved from the restaurant. They strolled over to her table and Jan asked Cincinnati, "Can I give you a kiss on the cheek?"

"Why sure, what for?"

As Cincinnati leaned down for his peck, Jan whispered that after she and David went to bed, they could hear Cincinnati and his Sportster Mob outside their window. Peeking out from behind the curtain to the patio area, they could see Cincinnati, a bearded and long-haired leaping gnome, running across a sea of titties, freaking and frolicking with reckless abandon. After that, Jan says, *they* jumped back into bed and enjoyed one of the best nights they'd ever had, Benbow Inn or elsewhere.

The Saga of Repo Jim

Every year around Eastertime a giant barbecue is held at Lake Worth, just about twenty miles outside Fort Worth. Live rock 'n' roll bands play by the water with Texas barbecue aplenty. Then there's the annual rattlesnake show, a truly unique Texas phenomenon where a foolhardy few carry on with live poisonous snakes. They swing 'em around by their tails. They dance with them. They play with them inside sleeping bags. Some contestants even kiss their pet rattlers.

Animal had seen enough.

"To hell with this snake stuff." Animal left the contest to rejoin his Jack County, Texas, MC brothers, the Rebel Souls. Out by the bikes, the boys burned a fat one.

LEFT TO RIGHT: *Animal, Repo's old lady, and Repo Jim hours before Jim's untimely demise.* (Photograph courtesy of Ed "Animal" Cargill)

As Animal and the brothers passed around the goods, Repo Jim took an extra-huge drag. He was in high spirits even without the weed, having just picked up a mint 1982 FLH Shovelhead a few days earlier. It was time to dump that Triumph he'd been riding.

Repo Jim gained his name and reputation as a repo man extraordinaire. It was what he did for a living, and he was damned good at it, too. Repoing was something you had to be half nuts to do in the first place. People pointed rifles, drew pistols, pulled knives, and did all sorts of crazy shit to avoid seeing their cars or pickups end up in the hands of a repo man. But Repo Jim was a patient stalker. He'd catch someone leaving his place, maybe to scoot down to the liquor store for a packet of smokes. Jim would follow him, jump inside his unattended vehicle, hot-wire the reins, and ride off. Another Rebel Soul club member named Taz occasionally rode shotgun with Repo, and he would strongly attest: When it came to the art of repo, Repo Jim was Van Go.

As the Lake Worth barbecue wound down, it was time for a ride over to the 2500 Club for an early nightcap. After a full day of riding, partying, and BBQ, the boys rolled up to the 2500 just before sundown and decided to chill out there for the rest of the night.

A few hours later, Repo Jim and his old lady decided to head back home. Repo's sidekick Bubba, who rode down to the 2500 in his truck, was splitting with him. Since the two guys lived close by, Bubba and his son offered to follow Repo home in the cage to make sure his new Shovel made it back to the house okay. Animal and the other brothers sent them off and continued partying.

A couple hours later, Bubba's son ran back into the bar frantic and out of breath. A bad scene had just gone down. Apparently Repo's Shovelhead had broken down on the highway (imagine that!). Since they had been following in the truck, it was decided that Bubba's son and Repo's old lady would stay with the bike while Bubba and Jim drove back to the 2500 to get help and grab some spare parts. But Bubba and Repo Jim didn't get far—a few blocks, in fact—when they got T-boned by a nineteen-year-old kid in a borrowed car. No insurance, no driver's license—the whole nine yards. Bubba walked away from the accident okay, but the wreck put Repo Jim in the hospital. Bubba's son had run all the way back to the bar to alert Animal and the other members about what had gone down.

Animal rounded up the troops for an emergency run to the county hospital. Once there, they held an all-night vigil at the ER to see that Repo Jim was okay. The county hospital was basically into patchin' 'em up and movin' 'em out as quickly as possible. That included drug ODs, knife wounds, car wrecks, and assorted contusions that made up a typical Sunday-night crowd.

It wasn't until sunup that Repo Jim staggered out of the emergency ward on his own steam. The remaining pack of Rebel Souls gunned their Harleys in the ER parking lot and beelined it back to Repo Jim's to watch over his recovery.

Repo was achy and sore—and barely coherent—as Animal and the boys sat with him the entire day. It wasn't until he fell into a peaceful sleep that they left his side. Early the next morning, Animal got an urgent phone call from Repo Jim's old lady. Trying to revive him and give him his medicine, she found poor Repo Jim stiff as a board. He never woke up.

And that's when the scramble began. Funeral arrangements had to be made, but Repo had no insurance. All his spare cash went into the Shovelhead. The club was his family. Animal and the boys, short on cash themselves, had to figure out a way to give Repo Jim a proper and honorable MC burial.

Animal contacted a longtime buddy, a smooth operator named Cowtown, who owned an old-style chopper shop down by the stockyards in Fort Worth. Cowtown had an idea. A few weeks back he had worked on a couple of bikes that belonged to the Canine Drug Unit. In trade for working on the cops' bikes, they paid him off in coffins they had recently seized in a drug raid. Cowtown had sold out all the coffins and most were paid for, except one. Cowtown's idea was that he would repossess the coffin, then donate it to the club.

And that's how it all came together. Five days after his untimely demise, Repo Jim's MC brothers had the situation firmly under control. Repo's mother already had a plot. The 82nd Airborne, Jim's regiment in Vietnam, contributed the flag, a ceremony, and a gun salute. Six other outlaw clubs showed up to pay their respects. In a dented black and chrome coffin seized by the cops and repossessed by bikers, and to the musical strains of Hendrix's "Purple Haze," Repo Jim was buried in his colors and his 82nd Airborne wings. It was an altogether fitting tribute to a fallen Rebel Soul, one of the greatest repo men in Texas. Ride on forever, Repo Jim.

DIY/DOA

DOA got his nickname for being a nonstop party animal, the kind of guy who *showed up* in the same fucked-up state that most people *leave* a wild party in. Other bikers called him "Dead Man Walking" because DOA had been in fourteen wrecks since he started riding twenty years earlier. To be fair, only two of the fourteen highway mishaps were on bikes; the rest were automobile pileups. Still, with that kind of voodoo hovering, people thought twice before stepping into a car with DOA.

When it was time to scoot, DOA had a menu of three Harleys to choose from. For short bar hops, there was the 1973 stroked-out Sporty. Next was another Sportster, a 1983 XLX. His A-rotation bike for long hauls was a powder blue 1990 Harley FLHS Electra Glide 1340cc V-Twin.

DOA on the road in one piece. *(Photograph courtesy of R. Scott Edwards)*

This is the story of DOA's worst bike accident. Fortunately (and as is common statistically), it occurred less than a mile from his house. It was a nice, biting, wintry Pearl Harbor Day in Pennsylvania. DOA tumbled out of bed, walked outside, and immediately knew that it was a nice day to ride. Never mind that he had just come back the day before from a solo run to the Gettysburg battlefield. He'd made that run in less than four hours from his home, a personal best.

While putt-ing through Pennsylvania Dutch country on the way to Gettysburg, you need to look out for a couple of things. First DOA Safety Warning: Be careful of those antique carriages that the Amish still use to get around in. You don't wanna be hauling ass around some twist in the road and smash into a horse and buggy's ass. That could get pretty messy. Second, you gotta watch out for those road apples littered around country roads. One false move on a chopper, and you'll end up losing some valuable traction on a pile of dung, fucking yourself up pretty good.

This one particular day, however, DOA's girlfriend was pissed off (as usual) at him for running off the day before. In an attempt to make peace, he promised to take her out to one of those cut-it-yourself Christmas tree farms to pick up a tree for the holidays. DOA's good buddy Duke and his old lady promised to come by with their pickup truck. They could all go out together.

That meant DOA still had a few minutes to kill, so he couldn't resist another quick scoot in the dry winter air.

He strolled out to his shed and rolled out his ride of the day, the '83 Harley XLX. As he walked into the house to grab his leathers and lid, his old lady began busting his butt again. "C'mon, DOA. Don't go out. When you get involved with that bike, you can be gone for hours."

"Forget about it," DOA told her. "I'm just going to take her down to the town square and turn back. No worries, okay?" Of course, she'd heard that line before.

The '83 Sporty fired right up in the thirty-degree air. Man, there's something about running a bike in the cold chill of morning that agrees with the compression of the motor. Nineteen eighty-three was the first year Harley dropped an EVO motor into its Sportster models, and this sucker ran like a top. DOA screamed out of his driveway and popped the clutch, zero-to-sixty in seconds flat. The torque harmony between the valves and the exhaust was almost musical. What a fine day to have the cold air slapping you across the face.

DOA was roaring, not even a mile away from his house, when he passed the construction equipment from a housing development being built nearby. Over the past few days, the trucks had laid a sheet of gravel and pebbles all over the road, and DOA hit a patch, rounding the corner at about sixty-five.

It all happened in a few seconds, but it felt like hours.

DOA skidded sideways on the bike. While he thought he could straighten course, by the time he got the bike straight, the centrifugal force of the skid had pushed him right off the main pavement, toward the shoulder of the road. As the XLX swerved on the bed of dry gravel and stones, DOA hit the soft embankment. Once the bike caught a mound of moist dirt, the Sportster twisted DOA off the bike and threw him like a bucking bronco, clean over the front handlebars and back out onto the highway. DOA was bodysurfing on the road as his helmet bounced like a rubber ball along the highway. Since he didn't have his leather zipped up (dumb move), he felt the horrible pain as he scraped across the pavement. Responding in a split second, DOA managed to roll over onto his back. At least he was dragging across the road on hide instead of skin.

Had DOA landed on his chin or his neck instead of his chest, he could have been killed or paralyzed. But hitting the road wasn't the worst part. Somehow, as he slid ahead of the bike, the wrecked Sportster was quickly catching up from behind. When DOA finally came to a stop, he then saw the bike coming right at him, like on a bad TV show. DOA watched helplessly from the ground as the Sporty bounced up and down like a pogo stick, up on its rear tire, heading straight for him! The bike bounced a couple more times, landing right on top of him, smashing DOA across the ribs, the frame lying perpendicular over his ribs and legs. With his breath knocked out, DOA lay there for a minute, barely conscious, trapped underneath the bike. He felt the hot exhaust pipes eating through his blue jeans, burning his shins.

As his adrenaline pumped, DOA lifted the bike up with one arm, like Hercules, and dragged himself from beneath the twisted Harley frame. DOA staggered to his feet like a sacked quarterback and shook his head. A shocked group of onlookers along the road pointed in disbelief, astonished he wasn't dead. But DOA was in tatters, bleeding, with gravel embedded in his bloody chest, completely road-rashed. His left arm dangled from his aching body. He felt (and looked) like Frankenstein's monster. An older man ran up to him. "Are you okay?"

DOA, still reeling from the accident, *thought* he was okay, but judging from the horrified look on the old fella's face, he wasn't too sure. His first thought was, could he possibly ride the Sportster home? His left wrist and hand were so badly fractured, he couldn't hold in the clutch. With his ride completely totaled, the walking wounded DOA left his prized XLX by the side of the road and limped the mile or so back home.

DOA's old lady was pacing out in front of the house, mad as a snake.

"So where's the bike?"

"I fucking wrecked it," DOA said as he wiped blood off his burning, sweaty forehead. "Anything else you want to know, like am I okay?"

"Does this mean we're not looking for Christmas trees today?"

That was it. Like a mad dog, DOA began ranting and raving, pounding his scraped helmet against the brick front of his house. Just then, Duke and his wife pulled up in the pickup truck. One look at DOA, and Duke suggested rushing him to a hospital.

"No fucking way! We ain't going to no hospital, no way," DOA winced as he shouted. "I don't want to hear for the fourth time today about this fucking Christmas tree. Let's go."

Jumping into the pickup truck, Duke helped DOA load his dead bike into the back and take it back to the house. Then everybody rode out to the Christmas tree farm, where DOA was a walking portrait in horror. He was still bloody, his clothing ripped and torn. Children ran in terror at the sight of this road-rashed zombie, burying their tearful faces into the pant legs of their fathers.

DOA was in throbbing pain and nearly in shock; he could hardly breathe. To add insult to injury, it was excruciatingly tough to cut down a Christmas tree with broken ribs and a busted wrist. After dropping off the tree at home, Brother Duke needed to convince DOA to head out to the ER. He had the perfect angle to play.

"Tell you what, DOA," suggested Duke, "let's go visit our buddy Zoomer in the hospital."

Zoomer was a straight-up biker—a Triumph rider and a country boy—who landed in the hospital after a horse kicked him in the heart, cracking most of his ribs. They had Zoomer wired up pretty good, and because of his heart, he was under constant observation, flat on his back.

As Duke and DOA walked into the hospital room, Zoomer took one look at poor, pitiful DOA and burst out laughing. In fact, Zoomer laughed so hard he grabbed for his chest, setting off all the alarms and monitors wired to his body. A SWAT team of nurses stormed into Zoomer's room. That did it. The head nurse finally had her fill of Zoomer's biker friends coming around and fucking things up.

"Both of you," she shouted, pointing toward the door, "out of here!"

DOA limped down the hall as Duke steered him closer toward the emergency ward. As they entered the ER zone, there was one nurse and one administrative assistant on duty with a roomful of flu victims. Being a poor (but incredibly cagey) biker, DOA tried handing over Zoomer's HMO number, explaining that his ribs were broken, he couldn't breathe, and he was suffering from chronic chest pains.

According to the nurse, since DOA was technically conscious and his bleeding had stopped, they couldn't treat DOA as an emergency case, especially since he didn't seem to have any noticeable means of insurance.

"Of course I'm not fucking bleeding," mumbled DOA under his breath. "The blood congealed after two hours of Christmas tree shopping in thirty-degree weather."

"Well, sir, you seem healthy enough to us," snarled the nurse.

And that did it. It was the end of a shitty day. How much worse could things get? DOA decided to take matters into his own hands. He leaned up close against an adjacent office door.

"Do you mind?" he asked sarcastically.

The nurse seemed confused.

DOA then grabbed his busted-up left hand, wedging it into the doorframe. Putting his foot up against the bottom of the door and shifting backward with his body, he then slammed the door on his wrist, applying enough weight to snap his wrist back into place. The "procedure" let out a horrible sound.

"CRRRRRACK." Frankenstein's monster was setting his own broken bones on the fly.

This shocked the assistant so much that she puked into the wastebasket next to her desk. The nurse ran off to grab security.

"Man oh man, that feels a whole lot better," sneered DOA.

DOA then grabbed a roll of packing tape off the nurse's desk and crafted himself a nice little cast for his sprained arm and broken wrist. After that, security hustled him out the front door of the emergency ward.

After a week or so of self-medicating, DOA never did reset the fracture. The road rash finally healed. The kids loved the Christmas tree, and DOA's old lady was off his case. Today his left wrist still has a funny little bend to it. But, hey, DOA saved himself some major medical bills while at the same time living up to his nickname. Do It Yourself, Dead On Arrival. DIY, DOA.

Behind the Locked Door

Deadwood is your classic Black Hills western town, situated slightly west of Sturgis, South Dakota, just off Highway 14A. Deadwood is steeped in Wild West heritage, in a lot of ways not having changed much since George Custer and Wild Bill Hickok walked its sodden streets during the late 1870s. Originally the U.S. government made a treaty with the Indians in 1868 that supposedly guaranteed their rights to the Black Hills. But when gold was found in the hills, it was futile to keep the white prospectors out of the area. Colorful talent like Martha "Calamity Jane" Canary, Annie D. Tallent, Poker Alice Tubbs, and Potato Creek Johnny gave Deadwood its color and reputation, particularly those who regularly robbed coaches leaving Dead-wood with gold shipments.

Deadwood's most famous casualty, Wild Bill, was murdered by a gunman at the Number One Saloon as part of a plot to keep him from being appointed town marshal. His presence in the barrooms threatened the town's outlaw population. He was shot in the back of the head at close range playing poker with a pair of aces and a pair of eights, later to be known as the Dead Man's Hand.

If you've ever been to the Sturgis Run and have taken a side trip to Deadwood, then you are probably aware of the fact that even though it prides itself as a real Wild West town, you cannot park your bike on the main street. You can either park in a hole next to a saloon or in a conventional parking lot.

Ever wonder why?

About twenty or so of us 1%ers, guys from all over America, were working our way back to California from one of our USA Runs. We came rolling into Deadwood about dusk, grabbed some rooms on the edge of town, and fanned out to check out the tourist sites.

This naturally led us to one of the local bars, the Famous Saloon, where Wild Bill also hung out, drank, and played cards. Come 11:00 P.M., some of our boys decided to break it up early and head off to our rooms to call it a night. We had ridden a whole lot of countryside that day.

Just as he dozed off, Cincinnati got the call. It was Griz from Connecticut on the horn. There had been a problem at the Famous Saloon with Train, RJ, and Dru. All three of our guys ended up in jail. Cincinnati pulled on some clothes and headed back to the Famous Saloon, but not before grabbing Young Gil and Griz out of their rooms for a little backup. As Cincinnati and the boys pulled up, people were still hanging around the Famous Saloon, standing out on the sidewalk. The cops were inside, and they would not answer any questions or let anyone back inside the saloon.

Just about then, a bartender came out with the garbage and Cincinnati asked him for a word. What the hell happened? According to the barman, Train was down in the cellar storage area of the bar getting his dick sucked by some local sweetheart. A youngster who did the cleaning up ventured downstairs, spotted Train and his sweetie doing the dirty deed, and ran back upstairs screaming bloody murder.

Now, the paragon of virtue who was blowing Train didn't want her rep to suffer none, so she yelled at Train to stop the kid. Train chased him upstairs and trapped him behind the bar. Someone threw an arm to stop Train. Train knocked him down. Some other fool came charging in and RJ knocked *him* cold. Then Dru reached over the bar and smacked another local across the mouth.

So a few people got smacked and some empty whiskey bottles broke. Nothing but pride and trash. Cincinnati asked the bartender if he would take him, Gil, and Griz back inside the saloon with him. Sitting inside at the bar, the local lawman was having a word with the owner of the tavern, and by the look of their shifty movements, these guys seemed to be up to no good. The cop finally left and Cincinnati talked a little turkey with the owner. Cincinnati immediately realized that this guy was a Grade A jerk, an opportunist. He claimed a whole bunch of stuff was damaged, but what it all finally boiled down to was eight hundred dollars. Pay up, keep your mouth shut, and the owner wouldn't file a complaint.

Deal, said Cincinnati.

Next Cincinnati, Gil, and Griz went hunting for the jailhouse. A few doors down, they spotted a building all lit up, but the door was locked. Coming down the stairs, they were met by a woman with brooms and mops.

"Is this the jail?"

The woman, a cleaning lady, said, "No, this is a whorehouse. The jail is down the street."

"Did somebody say whorehouse?" That got Cincinnati and the boys' attention.

The boys then headed off toward the jailhouse. Inside, behind the desk, was the same copper who had answered the complaint at the Famous Saloon. Seeing as how Cincinnati and he had been hard-eyeing each other in the barroom mirror, this wasn't going to be easy.

"I understand you have three of my pals locked up. I'm here to inquire as to the bail schedule."

The cop went on to explain that there was no bail. Cincinnati could hear Dru in the back of the jail hollering his ass off.

"You fucking punks think you're dealing with a bunch of country bumpkins," screamed Dru. "Well, you're the fucking hillbillies around here. I'll have more writs filed on your ass than you can believe." Dru was out of control. You could just picture him hanging on the bars, doing his best Jimmy Cagney.

Cincinnati asked the cop if it was possible for him to walk back and get Dru to shut the fuck up.

"Please," deadpanned the cop.

Cincinnati made his way back to the jail cells and there was Dru, relaxed in full repose, lying flat on his bunk, hands behind his head, only he was hollering up a storm. Seeing Cincinnati, he stopped.

"Hey, brother, what you up to?" Dru asked.

"Trying to get this fool to set you guys some bail."

"You think maybe I should be quiet, then?"

Cincinnati scratched his head and thought for a minute. Why should he? It was looking like Dru was really getting to the guy.

"Nah, don't bother. Keep it up."

"Cool," Dru said, and went back to his hollering.

The cop looked disappointed when he saw Cincinnati come back out, while Dru still carried on. Finally it became obvious that the cop was plain scared, in over his head, at least until the judge came in from nearby Belle Fourche in the next day or so to set bail.

Cincinnati asked for the judge's phone number. The cop looked back incredulously.

"It's one in the morning. What makes you think I'd be stupid enough to give you the number?"

Cincinnati convinced him that it was his intention to call the judge in the morning to help straighten this whole mess out. Surprisingly, the cop handed him the number, and Cincinnati and the boys beelined it straight to the nearest pay phone and dialed the judge's number.

"Ah, yes, Your Honor. We got your number from the deputy here in Deadwood, and he recommended we call you."

The sleepy judge cursed the stupid cop.

"What the hell . . . that dumb SOB . . . he knows there's a bail schedule . . . he knows goddamned well there's court tomorrow morning, ten A.M."

Back at the jail, Cincinnati told the cop, "Hey, Officer, the judge wants you to call him right now."

"But you said you weren't going to call him."

"What can I say?" Cincinnati said as he shrugged his shoulders. "I musta lied to you."

The cop dialed the judge, and there was all kinds of hemming and hawing on his end of the phone line.

"Yes, sir. No, sir. Never again, sir. Okay, sir." Then, in a whine, "Yes, Your Honor."

The cop hung up the phone with tears in his eyes.

"Hundred dollars apiece." Not a happy boy.

After having to strap Train and RJ on the back of two bikes, and getting Dru to hang on, Cincinnati got them all back to the motel and poured them into their beds. Then he and Griz immediately raced back to the whorehouse. Up the stairs they ran, knock, knock, knocking on the door.

A sleepy madam answered the door. Most of the girls were asleep, she said. She might have to wake them, so there might be a little noise. All of a sudden the soft-spoken madam turned into an angry drill sergeant.

"Clang, clang, clang, all right, you fucking cunts, let's unass them fucking beds. We got customers."

Cincinnati took the girl with the longest legs and headed for their room. This joint was straight out of the 1800s, a water pitcher and a bowl to wash your pecker. Just like the movies. After they'd done their thing, the girl informed Cincinnati that when Wild Bill was shot, they brought him upstairs and put him in this very room. Over there, she said, pointing toward the bed, was where he died.

After waiting for Griz to polish off his ladies, Cincinnati headed back with him to the motel. The next morning they assembled the rest of the crew (including the previous night's jailbirds) and headed off to the cemetery to check out the graves of Wild Bill Hickok and Calamity Jane.

Deadwood sits on the bottom of a very narrow canyon, while the cemetery sits on the side of a hill. Coming down off the hillside, roaring their straight pipes, knowing they were waking up the whole darned town (Deadwood: population 1,835), Cincinnati remembered the owner of the Famous Saloon waiting for his eight hundred bucks.

They turned onto the main street, where a group of townspeople stood on the corner, up the street, in front of the café where Cincinnati had arranged to meet Mr. Phony Fuck. Dru, RJ, and Train were riding up front, leading the parade. As the pack got closer, the fool realized that the boys were already out of jail and he was getting nada, zilch. He pitched a fit, angrily throwing his hat into the dust. On the next corner, the whores were hanging out of all the windows, waving and hollering out Cincinnati's and Griz's names to the wind. Cincinnati's old lady asks, "Cin, what are they saying?"

"Never mind, girl."

Later, it was found out that the bar owner had tremendous pull in the town. After this particular incident, he had developed a keen dislike for 1%ers, forcing a vindictive no-parking ban on all motorcycles. Absolutely no bikes are allowed at rest on the main drag, and that, ladies and gents, is why, to this day, you can never park your ride on the main street in Deadwood.

When Sara
Met Ron

Sara, a design school grad from Boston, loves to ride. Ron, a bike builder who lives in Pittsburgh, is an expert painter. He likes to burn tire. The two met in Laconia at Bike Week when they discovered they both rode bikes painted black and orange. At first Ron's buddies gave him a sound ribbing for having the same colors as Sara's '96 Sportster 883 (bored out to a 1200), a girl's bike. But over the week's festivities, Sara became friends with Ron and his crew. Sara noticed Ron and his boys rode some pretty unusual motorcycles.

"This guy builds some cool shit," thought Sara, making a mental note.

Ron's scooter was a 1973 Ironhead rigid-framed chopper. Parked next to that was another

🔥 **Sara in a calendar shoot with a couple of Ron's customized choppers.**
(Photograph courtesy of Jim Nocera)

Ironhead, this one with a Hardtail swing arm. Then there was a 1986 SSL converted into a chopped-down Road King. Alongside that sat a beautifully restored 1980 Shovelhead inside a rigid frame. Ron built and painted the entire fleet of bikes.

Sara grew up the daughter of a steel-working man, Daddy's favorite girl. She was heavy into metal, poking around her dad's steel business, looking for odd pieces to weld. Dad was also a hot rod builder and an antique car buff, so when young Sara got her driver's license, father and daughter rebuilt a '66 Mustang for her to tool around town in. Installing a new carburetor,

Sara learned a little about wrenching. When she entered design college in Massachusetts, Sara integrated steel and welding into the scheme of her photo artwork. While at school, Sara started hanging out with some of the local New England bikers, which is where she learned to ride motorcycles.

Boston is an expensive city to start out in, especially if you consider yourself a struggling artist. After graduating from the design academy, Sara freelanced as a photographer for a few magazines. At that time, motorcycles and photography represented two separate worlds to Sara. It wasn't until she ran into a bike mechanic named Mark that the concept of motorcycles and art finally collided.

Mark, like Ron, was a quiet biker, a loner at heart. He liked to work alone in his small, one-bay bike shop in Methuen, Massachusetts, next to the New Hampshire border. Sara met Mark through a mutual riding friend, and when it came to custom cycles, underneath the traditional biker trappings of long hair and beard, Sara could see that Mark was a genuine artist. He rode a '48 Panhead that he'd transformed into a cool 1960s retro chopper with high-set rear pegs and a big king and queen seat. ("One of the baddest bikes I'd ever ridden on," says Sara.) On the outside Mark looked like a mean guy, but underneath he was a helpful sort. Pretty soon the two struck a deal. If she kept out of his way, Mark would let Sara hang out in the shop, let her take her damned pictures while Mark wrenched his day away in relative peace.

"I'd photograph bikes, engines, parts, tools, whatever caught my eye. My camera equipment took up a lot of room, but fortunately Mark didn't seem to care."

After shooting countless rolls in Mark's tiny garage, Sara finally found her photographic inspiration, mixing body parts with bike mechanisms. Pistons and cylinders merged with the curvature of the female anatomy. Spooky *Terminator*-type stuff connected through common tone and lines.

"Motorcycles were designed by men specifically for men," said

Sara's self-portrait using her trusty Nikon.
(Photograph by Sara Laliberte)

Sara. "You throttle with your hand while you shift with your feet. A bike is hard, rough, cold, and tough. So I started to look for ways to connect that with the female body, which is the opposite, soft and curvy."

Mark was so impressed with the photos she took that he encouraged Sara to send her stuff to some of the bike magazines that printed experimental photos. *In the Wind* had a section called "500 No Shit," meaning if you sent in a cool photo and they printed it, they'd mail you five bills. No shit.

One day Sara's phone rang. It was a mutual friend calling about Mark. He had just finished a bike, one Sara distinctly remembered him working on. She had taken several photos because of the tank design. Mark took the bike for a final spin, but out on the street, an accident with an elderly driver terminated the test drive, and Mark died in the ambulance on the way to the hospital. Sara was stunned. Just days earlier, Mark was happy with life, wrenching his bike in his single-bay paradise. Mark had a typical biker's big heart, always opening his doors to help out other riders just as he had done for Sara, stopping on the highway to assist when he could. Now he was gone.

Artist at work:
Ron painting Sara's designs
on saddlebags.
(Photograph by Sara Laliberte)

"2 Into 7"—a photo
collage by Sara Laliberte.

When *Bike Week was over in Laconia, Sara and Ron* exchanged phone numbers and kept in touch. After one visit to Pittsburgh to see the Andy Warhol Museum, Sara relocated to the famous Steel Town. Pittsburgh is a major biking city. Weekend pig roasts at the Voodoo Lounge attract up to five or six thousand bike riders. As for the riding, one minute you can cruise city streets, twenty minutes later you can ride the open western Pennsylvania countryside.

Ron's customized bikes have gained him a reputation among Pittsburgh riders. Sara solidified her work, merging collage art with her love of motorcycles. Buying out a local shop, they opened their own small storefront with a large garage ideal for welding, fabricating, and paintwork. The two are together nonstop, living, working, and breathing the same world of motorcycles. Word of mouth has continued to generate business from as far away as Cleveland.

"We're such a great team, it's scary."

Eventually *In the Wind* magazine printed Sara's photo of Mark's last ride. It was a full-page image in the magazine captioned "In memory of Mark, ride in peace."

Meanwhile, Ron and Sara forge ahead with their custom work and designs. Ron perfects his painting technique and burns tire whenever he can. As for Sara, she found her groove mixing motors and photos.

The Ballad of Rocky's Green Gables

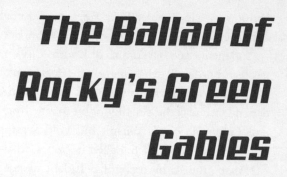

Near East Truman and Winner Roads, about four miles from the town square of Independence, Missouri, stands a wooden tavern with matching green gables. After its early days as a dance hall, Rocky's Green Gables built its reputation as a biker bar. It was known far and wide. Besides catering to the bikers and working-class neighbors, Rocky's regulars included workers from the nearby oil refinery. According to the *Kansas City Star*, Rocky's Green Gables held the second-oldest liquor license in all of Jackson County. It was the beer that kept the place rockin', but it was the bikers who gave Rocky's its personality. Some of the regulars believed it was their Green Gables that Waylon Jennings sang about on his 1973 country classic "Honky Tonk Heroes," although the song is actually about another Green Gables bar in Waco. But so what?

Independence is a suburb of Kansas City on the edge of Missouri's western border. It was where American explorers Lewis and Clark began the Western Expansion. As the "Queen City of the Trails," Independence was the starting point for the frontier Santa Fe, Oregon, and California Trails during the Wild West era. Mormon Church founder Joseph Smith Jr. called it Zion, God's own city on earth.

Most Americans recognize Independence as the birthplace of President Harry S. Truman, but it's also the infamous methamphetamine capital of the Midwest, if not the entire United States. Independence has always been a biker-friendly community, a recovering working-class town that has seen more than its share of manufacturing jobs go away.

Kid Jeremy was a young rider who grew up just down the road from Rocky's Green Gables. As a youngster on his bicycle, he and his pals watched the hubbub from a safe distance as bikers gunned their choppers in and out of Rocky's gravel parking lot. Growing up in a hot rod and motorcycle (pronounced "sickle") household, Jeremy was always big for his age. Born of healthy German stock, Jeremy now stands six-two, 230 pounds. He vividly remembers his first bike, that 1969 Triumph locked up in the backyard shed. Jeremy's father, John, was a respected mechanic, hot rod builder, and bike chopper in town. Father John rode with all kinds of biker types and was close friends with hot rod legend Ed "Big Daddy" Roth. Whenever Roth was in town for an auto show, Big Daddy would crash in the family guest room.

Almost as soon as Kid Jeremy learned to walk, he wanted to ride, and for as long as he could remember, the roads of Independence always echoed with the sound of motorcycle pipes.

Jeremy grew up in a yard littered with spare parts and go-carts. Father John rode mostly Triumphs but always kept a Harley or two handy, including a super-cool late-1960s Super Glide sporting a Willie G fiberglass back end with a '59 Caddie taillight. Most of the

Triumphs that passed through the yard ended up custom choppers, except for the virgin Bonneville Triumph that remained imprisoned in the shed.

The Kid kept his eye on the Bonneville. When he hit age eighteen, he made an aggressive play for the Triumph as his main ride. But John discouraged him, figuring he was still too young for the bike. He could hurt himself on the damned thing and wreck a perfectly good bike. But Jeremy refused to let the subject of his first real ride, ride.

He finally told his dad, "Look, I'm old enough to vote and kill, so I'm gonna buy a motorcycle." Father John finally caved and gave up the Bonneville Triumph. Jeremy rode the streets of Independence and kept his nose clean. When he turned twenty-one, Father John took him out for his first (legal) cold one at Rocky's Green Gables.

Riding up together on their bikes, Kid Jeremy and Father John could tell who was inside by the hogs parked outside. Twisting the knob of the green door, Jeremy's hand was actually shaking.

"This is gonna be good," Jeremy decided. "I'm part of the mob."

The long wooden bar was one of the oldest in the Kansas City area, and the beer was undoubtedly the coldest, even colder than the brews ordered up at other biker haunts like Frankie D's, the Other Place, and Faces Lounge. One of the first things a regular learned at Rocky's Green Gables was that the guy behind the bar running the joint wasn't named Rocky at all. Newcomers might call out to "Rocky" for another round, but a guy named Ron ran the place.

The jukebox was well stocked with enough Southern rock and C&W to choke Charlie Daniels. "Green Grass and High Tides" by the Outlaws was a favorite, along with plenty of Waylon, Willie, and Johnny Paycheck tunes. A huge Dixie flag hung over the jukebox.

Jeremy's first legal drink inside the barroom was like a scene lifted out of the movie *Goodfellas*. It was crowded with a cartoon cast of regulars. The first voice you heard was Mayor Ronnie, who actually did get elected mayor of Independence and scooted around town on a tricked-out, cobalt blue Road King. To his left was Danny, who rode a mean metal flake chopper that was a Sportster in a previous incarnation. Next was Mr. Big from Milwaukee. He had been around Harleys his whole life and went back to visit his hometown (with a bunch of Rocky regulars) to celebrate one of H-D's anniversaries in the biz.

Bones, a retired ironworker, rode old Triumph choppers for several years just like Father John. Guys like Bones knew where the bodies were buried in terms of fixing old motorcycles. Between Father John, Bones, and the others, Kid Jeremy would forever tap the Rocky brain pool in order to keep his own ride running tip-top.

Most of the Rocky regulars were like Jeremy's father, traditional motorcyclers, guys who worked on their own bikes and rode them hard. They were quite different from most of the younger riders Kid Jeremy knew, the ones who barely changed their own oil. Even though he had younger friends of his own, Jeremy had more in common with the old salts at Green Gables. They drank loads of beer, gunned their engines, and full-throttled life one day at a time.

In spite of its rowdy facade, Jeremy found Rocky's Green Gables the ultimate kickback hang, provided nobody stepped too far out of line or bothered anybody in need of a couple of cold Pabst Blue Ribbons and some solitude. There was rarely any trouble, only the occasional scuffle, certainly far less than the doorway skirmishes at the trendy dance nightclubs where the Kid sometimes bounced for spare cash.

As the two grew older, Kid Jeremy and Father John spent more and more time together riding. To the casual onlooker, they seemed more like pals than father and son. As the years passed, Father John gave up riding chopped Triumphs for a new pearly white Heritage Softail with a full windshield and tooled leather saddlebags. Jeremy, too, had developed a more basic taste in bikes when he shed the Bonneville for a Harley Super Glide. The two rode up north to Sturgis and camped out in the Black Hills together, freezing their asses off in the open air.

Like father, like son, like good buddies—Jeremy (LEFT) and Papa John.
(Photograph courtesy of Jeremy Povenmire)

"Jeez, we were a couple of Missouri guys. It was supposed to be August. What the hell did we know?"

Just before another trip to Sturgis, Kid Jeremy got married. Incorporated into the wedding vows for his new bride was the line "to love, honor, obey, and polish Jeremy's bike." He and his wife moved into his grandparents' old house in the same neighborhood as Father John. Alongside Jeremy's Super Glide sat his wife's Kawasaki Eliminator 600.

For Kid Jeremy, the nicest part of living in Independence was always the riding. When the Rocky regulars got together for an unofficial Green Gables run, they insisted on Kid Jeremy leading the pack, front left, with Father John riding front right.

All right, you Philistines! Everybody hit the deck and nobody move. This here is a stickup! Keep your faces to the floor!"

In 1960—the year Kid Jeremy was born—there was a low-rent bank robber from Independence named Guy who pulled off his first heist. Guy hooked up with a wheelman in Jefferson City, and both played the armed robbery game in Kansas City and fled to Southern California, where the pair continued knocking off more banks in the greater Los Angeles area. Guy was twenty-one when he was finally picked up and hauled in by the cops. Since he was the inside man—waving the gun and yelling out the orders—Guy was the only one identified on the lineup. When the cops pressured him to put the finger on his partner, Guy wouldn't budge.

"Back when I was a criminal, we didn't rat out our partners."

The state of Missouri soon caught up with Guy and his partner and extradited them back to Kansas City. Guy's partner got twenty-five years, but not before the wheelman's girlfriend smuggled a couple of hacksaw blades and files inside the county jail. She did two years herself for pulling off that crazy stunt and lost custody of her kids.

As for Guy, a cell in the Leavenworth federal pen would become his residence for nearly eight years. By 1968, he was back on the streets alone with neither a home nor a saleable skill. Motorcycling had always been a huge part of Guy's life since he was fourteen years old. His first bike was a Cushman, one he bought for fifty bucks off a carney who ran shows on 18th and Minnesota Avenue in Kansas City.

Released from Leavenworth with only a prison GED in his back pocket, Guy stumbled from one menial job to the next for five years, until 1973, when he landed a decent gig driving trucks and hauling automobiles. Surrounded by four-wheels of all kinds, Guy still loved motorcycles. He picked up a 1959 Panhead for three hundred bucks. It was full-dressed and loaded up when he first got it; by the time he took it out on the streets it was stripped and chopped.

Guy didn't ride with any one club during the 1970s, although he did hook up with several of the local KC 1%ers. At the time Guy was a loner, not a joiner; he couldn't handle a three-piece suit, much less a three-piece patch.

The 1970s proved to be a blur for Guy, and when the 1980s came around, he was a bummed-out ex-con. Then Guy met a rock-stable woman named Rhonda. With a two-year-old, Rhonda knew about responsibility. Soon the two were married. A year later, they had a child of their own and became a family of four.

Rhonda was a Christian; Guy was mixed up as hell, pissed off most of the time, and now had to provide for a family. But except for riding his motorcycle, nothing much mattered to Guy. Eventually a buddy at work invited him to his church. Within a short period of time Guy "accepted the Lord" at the end of the service, and took that fateful walk down the aisle to the front of the church.

Guy tried to live "the Christian life"—shave, get a haircut, buy a suit, and become part of the mainstream. But riding his motorcycle as much as he did, Guy just didn't fit in. He attended Christian Motorcycle Association rallies, but most of the CMA guys were just weekend riders.

Then Guy ran into a biker nicknamed Brother Bill the Baptist hanging out in front of the CMA rally. For small talk, Guy commented on Bill's ride, a Super Glide with drag bars and a narrow glide front end. He explained his situation to Brother Bill. For a long time Guy struggled to reconcile some kind of connection between his own biker lifestyle and his newfound religion.

Guy confessed to Bill at length about the unexplained yearning he felt in his heart to pursue some kind of ministry, but he didn't know where to start. Bill understood his dilemma. Somebody had to reach out to the biker community. Sure, they were a rowdy bunch, but yes, they had heart and soul, and more often than not supported a stronger system of values and a sense of honor that burned brighter than your average churchgoer.

Brother Bill nodded and sympathized. Besides wearing a three-piece patch from a local club (God's Journeyman) in KC, Bill the Baptist also hosted his own weekly prayer meeting. Bill the Baptist and Guy took a long ride that night. From behind his high bars, Guy decided what he needed to do: Preach to the needs of the motorcycling community just like his new friend Bill the Baptist. How hard could it be?

Guy spoke with all of his close biker friends, the folks he drank and partied with before his conversion. But Guy made no headway. He was about as successful at winning souls as he had been at robbing banks. Guy bounced around the KC area on the cusp of making a name for himself. What eluded him was a firm direction. That

Guy Girratono preaches the Gospel on the road at a bike-rider event.
(Photograph courtesy of Guy Girratono)

is, until months later when Independence lost a beloved bike rider everybody called Redbeard.

Redbeard was running the corners out on Noland Road, a real curvy Independence highway, which he was prone to do. Redbeard loved to drag the running boards on his Heritage, only this time he miscued, went off the road, crashed and burned.

Redbeard's mother wanted a true biker's funeral, so about thirty or so bikers rode in the snow to Bolivar, 130 miles southeast of Independence. It was cold as blue blazes. At the service, a couple of Redbeard's good friends eulogized him. Big, old, rough-looking, bearded men wiped away tears of sadness. Death, said one friend, is a more common obstacle in the world of bike riding. It's the necessary risk of experiencing the code of the road. Just before the pack left the funeral home for the cemetery, they opened the casket, and there lay Redbeard, dressed in his leathers and his colors, his lifeless hands clutching a bottle of Southern Comfort whiskey. That did it. Another lost brother meant another lost opportunity for Guy. Why hadn't Guy gotten to Redbeard before he died? And if he had, maybe he wouldn't be clutching a whiskey bottle.

Out at the gravesite, watching all the sadness over losing a brother, it further dawned on Guy: Redbeard's death was the motivation Guy the Preacher needed to kick-start his own ministry to the bike riders of Independence. He needed to become a visible example, so in 1990, Guy staged a run called Bikers with a Mission to raise money for disadvantaged families. Not knowing who or how many might show up, he was happy to see nearly six hundred riders, ranging from the smallest scooters to the biggest V-8 boss hogs, rev up their bikes to raise $23,000 for the mission at the first run.

Over the past ten years, Guy and the Greater KC bikers raised over $300,000 for City Union Mission's Family Center. Guy, the former bank-robbin', chopper-ridin' preacher, had his own motorcycle ministry in full rev. Now all he needed was a permanent site for a church.

Guy the Preacher (along with others from the Disciple of Jesus Ministries) continues to scour the state of Missouri, tending to lost souls. With a small sound system and tent, they visit bike rallies across the state. They serve coffee to hungover bikers, feed people, and do some one-on-one prayer talk and testimony about the Lord, as well as preach the Gospel on Sunday mornings.

Guy gets mixed reactions at the biker rallies. His crowd can be quiet and attentive. Some come to score a meal or a cup of coffee. A few find the Lord. Others burn rubber and throw gravel as girls raise their tops in an effort to poke fun or drown out the message. But Guy is determined to preach and hell-bent on reaching the Harley crowd, from HOGs to 1%ers.

Ron the barkeep and his wife, Linda, seriously pondered the future of Rocky's Green Gables. For sixty years it had been a trusted local establishment, but the place was definitely showing its age. Originally built without indoor plumbing, the place had deteriorated into ragged shape. The plaster was old and cracked. No insulation. The hardwood floors, warped. The bar had leaked for so long that several layers of plywood replaced a proper tile floor.

One night Guy the Preacher got a call from Ron and Linda. Later Guy met with Ron and Linda at the bar. They spoke about the old times at the Green Gables and its colorful biker clientele. Ron and Linda had a proposition for Guy.

"Are you still in need of a building for your church?" Before Guy could answer, they asked, "Would the bar work?"

Rocky's was in rough shape, but eventually they struck a deal that worked for everyone involved. Guy's church got the bar for a reasonable price, while Ron and Linda got out of the bar business, avoiding a costly renovation. And over the next several months, Rocky's Green Gables morphed into the permanent home of the Disciple of Jesus Ministries. It was now time to gut the historic biker bar.

Ron the barkeep had already collected contributions from a collection jar he kept on the bar before closing it. Another biker, Old Mike, passed the hat for another seven hundred and fifty dollars.

Guy on his "bad boy" Grape Escape in front of the Disciple of Jesus Ministries, formerly Rocky's Green Gables in Independence, Missouri.
(Photograph courtesy of Guy Girratono)

Then an army of bikers teamed up and donated lumber, hammered up drywall, framed and plastered the inside, painted the place inside and out, and did all the electrical work. The Independence biker community kicked in a total of twenty-five grand to transform the old biker bar into a new biker hangout of a different stripe.

By early 1998, Guy preached his first sermon at Independence's first all-biker church.

Now when locals ask where Guy's church is, all he has to say is "Rocky's Green Gables." He rarely gives out an address. The rock of Guy's new biker church has more than solidified his spiritual foundation. Only his faith in Harleys has been shaken. Riding in on the Grape Escape, his candy purple and teal Honda Valkyrie with six carbs, six straight pipes, extended fenders, and high bars, Guy the Preacher has officiated at over one hundred biker weddings and

A *Rocky's relic: the remaining neon now living over the workbench inside Jeremy's garage.* (Photograph by Jeremy Povenmire)

probably half that many funerals, including some for the local out-law clubs. After a funeral, Guy might be invited back to the outlaw clubhouse for dinner and conversation.

"That just proves to me that we're accepted, and there are no problems."

Meanwhile, back down the road, with Rocky's Green Gables merely a memory, Kid Jeremy retires to his garage work-shop. He switches on the Green Gables neon sign that he rescued from the bar shortly before the sledgehammers fell on the place. The sign emits a soothing glow of green that bathes the en-tire garage. Raising a chilled Pabst Blue Ribbon beer, Kid Jeremy and Father John toast the bar and the bikers (alive and dead) who made Rocky's a one-of-a-kind place.

"Here's to Rocky's Green Gables," Kid Jeremy says, raising his Pabst high. "To the place and people who gave so much to us, the independent bikers of Independence."

Blake's World-Record Whorehouse Jump

Blake from SoCal has been ridin' high for the past four decades. He bought his first Harley at age fourteen and got hooked on motorcycles as a little kid when he lived near a fire station. Each day when one of the firemen rode to work on his Harley-Davidson, Blake would listen for the roar of the pipes, then run down and sit on the curb across the street from the station to admire the bike parked outside.

Just after his thirtieth birthday, Blake made friends with a wacky 1%er named Seth. Seth was a member of a Big Four club that operated out of the Great Northwest. Blake first met him and his partner, a guy named Jones, when the two walked into a Montana barbershop that Blake ran. It was the dead of winter, and Blake overheard the two discuss a run they were planning down to the Winter Nationals in Pomona, California. Blake, bumming through a messy divorce, saw a way out of Montana and shocked himself when he butted in on the two guys' conversation.

"I ride a Harley, too. You reckon you'd be interested in having a third person along to help share expenses?"

Seth looked over at Jones. "Well, stranger, we'd like to talk it over and maybe get back to you on that."

"Well, let me know," said Blake. "I sure as hell wouldn't mind trading in this Montana snow for a ride in the sun."

A couple of days later Seth walked back into the shop and gave Blake the go-ahead. That night, over a few brews, the three sat in the bar and plotted out the details of their trip. The next morning they took off for California, and from that day on, Blake never laid eyes on Montana again. Plus, it was the start of a long friendship between him and Seth.

Months after the Pomona run, Blake rode into the small Washington town where Seth hung his hat. Seth was still a member of the same 1%er club, which operated a "body-painting studio" on the outskirts of town. The scam was this: six or seven of the club's sexiest old ladies were "rented out" as "canvas" for group affairs or private "artistic" consultations. Seth was the manager/bouncer of this fine establishment, which danced a fine line between semilegal art appreciation and convivial carnality.

Seth was a notoriously fun-loving kind of guy. He wasn't particularly tough, and he wasn't particularly mean, nor was he particularly robust, either. But he sure could raise hell and fight dirty at a moment's notice. Seth had a habit of looking for trouble and pushing the wrong people around. Unless he was surrounded by a squad of rat-packin' brothers, Seth, more often than not, became the "stompee" rather than the "stomper." Still, for all the times Blake pulled Seth out of one close scrape after another, the trade-off was that Seth was the world's greatest hang. It might be a bar, maybe a tattoo parlor, wherever; Seth usually had an innate sense of where the fun was about to unfold next.

At the body-painting studio, there wasn't a whole hell of a lot for Seth to do other than keep one eye on the till and the other eye on the peace, in case a john—or an art enthusiast—felt gypped or ripped off. The day Blake stopped by, he found Seth hanging out by the back door, bored and shitfaced from a day's worth of drinking Southern Comfort. Surprisingly, Seth's "studio" was crawling with customers, an uncharacteristically busy day for the middle of the week. Seth was glad when Blake pulled up. He immediately asked for a favor.

"My ride's been acting up. Whaddya say the two of us drive your pickup over to my place so I can haul my wheels back up here to the studio? I have a spare room where I can fix it while I'm sitting around here doin' jack."

Seth and Blake drove to Seth's pad, picked up the motorcycle, and drove back to the studio. They pulled up near the rear entrance and rolled the wounded Harley off the pickup. There was a loading dock at the back door. All they needed were some wooden planks to build a ramp up to the platform where the back door was. Once they built the makeshift ramp, they noticed it was still kind of steep. Since Seth was still a little high, they decided that Blake should ride the Harley up the ramp. He'd get a good run at it, give the bike some gas, and stop quickly enough once he reached the top of the dock. Seth opened the back door to the storeroom just in case Blake needed some extra stopping room. Next to the storeroom were some of the "art studios."

Blake mounted the Harley, kick-started it, gave it some gas, and lined up a couple of yards behind the makeshift ramp. All of a sudden he realized that the throttle was stuck. The bike took off up the ramp like a shot. It sailed through the air, through the back door, through the storeroom, and crashed straight through the Sheetrock wall toward the other end of the parlor. Blake landed smack-dab in the middle of a "body-painting session," missing the girl and her trick by just two feet. The rest of the customers were scrambling to grab their clothes and run for cover. Blake, rolling on the floor, laughed like a madman. Seth came running in all bug-eyed.

"What the fuck happened, and what's so damned funny?"

Blake got up and dusted himself off. "Eat your heart out, Evel Knievel. I now hold the world's record for jumping a Harley into a whorehouse."

To this day, it's doubtful that record has been, or ever will be, beat.

Loaded Linda's Silent World

During the biker psychedelic 1960s, the streets of Oakland were paved with parties. A one-of-a-kind cast of characters roamed the Bay Area highways like modern bandits. Club guys like Pee Wee, Crash, Salem, Limey Joe, Tony the Wanderer, Freewheelin' Freddy, Garcia, and East Bay Jake were kicking ass, riding high, and livin' free. And so were a lot of young, eager-to-please chicks who hooked up with the Club. They may not have worn patches or fought alongside us in the bars, but they had front-row seats (usually from the back of the bike) watching it all go down.

One of the coolest women to hang out with us was a high school chick we dubbed Loaded Linda. Linda drove her parents batty. As teenagers, she and her good friend Deanna would cut school and camp inside the parked boxcars in the Richmond railway yard to smoke pot.

Linda started hanging out back in 1963. It was a helluva stretch for a rebellious high school chick from Richmond to go from begging for motorcycle rides from teenage hoods in the school parking lot to actually packing with the elite members on their way to Bass Lake runs.

Linda's first contact with the Nomads came when the guys moved down south, making Richmond their temporary base.

Linda knew Tony the Wanderer. At the time Tony lived with a girl named Lana. Lana was a hard-boiled broad and she was extremely street-smart. Lana turned tricks on the street to support herself and Tony. Naïve Linda was shocked to find out that she'd made her first acquaintance with a true-to-life prostitute.

Tony was well known in the Bay Area psychedelic scene. He fancied himself a player—half-1%er, half-hippie—and made friends with influential, turned-on hipsters like LSD pioneer Owsley Stanley, the writer Ken Kesey, and psychedelic guru Timothy Leary. All the top Frisco bands like the Grateful Dead, Big Brother and the Holding Company, and Blue Cheer rolled with Tony. While Lana turned her tricks, Tony entertained a different wide-eyed hippie chick on the back of his bike every week. If you needed to find him, chances are he'd be holding court at happenings like the be-ins in San Francisco's Golden Gate Park or at free concerts at the Panhandle in Frisco. It was there that Tony would deal acid and firm up his Haight-Ashbury druggie connections.

This was the crowd Linda found herself hanging with after she graduated from high school. Linda's very first ride on the back of a motorcycle was with her first boyfriend from the Club, a Nomad named Robbie Garcia. It wasn't exactly a good ride. They were making their way up a winding hill on Santa Rita Road in El Sobrante when Linda's foot slipped off the peg, hit the shifter, and kicked the bike out of gear. The bike rolled backward downhill. Robbie steered the high bars as best he could as the bike veered wildly left and right down the steep grade. Traffic whipped in and around the out-of-control two-wheeler. After several near misses they made it to the bottom of the hill. Robbie was plenty steamed when Linda asked why they ended up back where they started.

Loaded Linda learned to hang on as Robbie taught her to lean in on the turns. Once she took Robbie's advice a little bit too seriously and leaned in to the point where Robbie nearly had to lay his bike down on a busy highway. After two close calls, most guys would have knocked her right off the bike. But Robbie was a patient rider.

I *first met Loaded Linda in 1965 when a bunch of the Nomads* showed up at the El Adobe Bar on East 14th Street. Robbie introduced her as his old lady. She was only eighteen, flashing a fake ID to get into the bar. I was married to my first wife, Elsie, at the time, so Linda was relegated to the table with Elsie and some of the other gals who sat together smoking and drinking. They offered her some Seconal pills, which was Linda's introduction to taking reds. Trying to prove she could keep up with the other old ladies, she popped one after another until she passed out beneath the table. Robbie slung her over his shoulder Tarzan-style, carried Linda outside, and stashed her in the backseat of a car to sleep it off.

Loaded Linda was a pretty red-head. Her long thick hair would flap in the wind as she packed with

Lovely Loaded Linda at a Bass Lake run. *(Photograph courtesy of Linda Black)*

Garcia. But what made Loaded Linda so truly unique out of the hundreds and hundreds of other old ladies who hung with us over the years was that she was virtually deaf. Linda lost her hearing at age two after she caught the German measles. The disease destroyed the nerve endings of her inner ears, wiping out 90 percent of her hearing. Linda went to a speech therapist and learned early how to lip-read. Yet when Linda rode on the back of our bikes she could hear, however faint, her favorite sound—the roar of a pair of Harley-Davidson straight pipes. Hanging out with us was her way of showing her protective family (and, yes, herself) that she could make it on her own in spite of her near silent world.

Linda continued packing with the Nomads. One night the whole chapter met up and rode into San Francisco together to see Janis Joplin and Big Brother and the Holding Company headline the Fillmore Auditorium. Linda rode in with Limey Joe, her new boyfriend from the Nomads. An army of bikes roared up to converge on the entrance at Fillmore and Geary. Bill Graham, the hard-boiled concert promoter, was working the door himself that night, and he took one look at the pack and their chicks and freaked out. If they wanted to get in to see the show, as far as he was concerned, then they could fucking pay. This created a potentially bad scene out in front of the hall, since Janis herself had promised to take care of the boys on her guest list.

Somebody threw down some cash and bought one ticket for Linda, who made her way to the backstage area, where she found Janis, resplendent in her usual array of feather boas and beads.

"Janis, there's a whole bunch of bikers outside to see you and Bill won't let us in."

"We'll see about that."

Apparently Graham had been messing with Janis's guest list all night and this was the last straw. Janis refused to go onstage until Graham opened the doors and let in all the guys and their old ladies. From that night on, even when we weren't on the guest list

The Nomads from 1970. Limey Joe is second from the right.
(Photograph courtesy of Linda Black)

for bands like Sly Stone, the Grateful Dead (who welcomed us anyway), B. B. King, or the Who, we never had to pay.

Linda's favorite part of the night was the ride back from Frisco
to the East Bay. After the concert, a pack of two dozen or so Harleys
kick-started their bikes for the fast ride home. Some of the guys
were laying down some last-minute sweet talk to a few Flower
Power blondes with miniskirts and hair down to their asses. As the
pack gunned their engines, there were always two or three stragglers too loaded to start up their bikes. Eventually everybody was
ready to roll. Then a collective, thunderous barrage of rumbling
torque sliced through the San Francisco night. The block-long pack
roared away from the Fillmore, up the Geary Street hill, then made

a quick right downhill on Gough, beelining it straight for the Bay Bridge. The bars were closed. The Frisco city streets were nearly deserted. There was no need to stop for stoplights; just roll on through the night 'til you hit East Bay grease.

Riding through the tunnel that connects both parts of the Bay Bridge heading east was the biggest thrill for Loaded Linda. There was nothing she could hear better in her muted world than the full-throttle, ricocheted sounds of twenty-five pairs of Harley-Davidson straight pipes bouncing cacophonously throughout the short stretch of tunnel. It was a major high riding with Limey Joe. Loaded Linda held on tight and felt on top of the world. With the roar of the motorcycles and the pipes revving full blast, everyone felt the unbridled power and propulsion of their bikes.

Tony was way in the zone and digging things, too. He had broken ahead of the pack, but with a renewed head of steam, the rest were catching up fast. As he sped through the tunnel, Tony veered back and forth across all five lanes like a drunken, freewheelin' warthog. Then he leaned way back on his sissy bar, stuck both feet up high in the air, and howled like a lone wolf. The raging horde zoomed out of the tunnel like a swarm of angry hornets—past Treasure Island, off the bridge, and into the open air. It was there that the pack would separate. Nomads veered left toward Richmond, Berkeley, and points north. The rest of the riders split off in the opposite direction, toward the Nimitz Freeway to Oaktown, Hayward, and places south.

Somewhere just off the Bay Bridge, Limey Joe (packing Linda) and Sergio (packing his wife, Anna) began racing. They weaved in and out of lanes, dodging clueless drunk drivers, hitting speeds of over 100 mph. Linda worried about how much dope she and Limey had taken. He had dropped a couple of tabs of Purple Haze LSD before Janis hit the stage, and swallowed a handful of reds to come down after the show. Freaking out from *her* acid tab, Linda felt the rear tire slip and slide a couple times as Limey closed in on Sergio, who was leading by a good length and a half.

Linda felt the back of the bike slide and skid for a third time. Then she did the chicken thing.

"CHP on the ramp!" Linda shouted into Limey's ear.

Limey rolled back on the throttle and slowed down quick. With another speeding ticket, he risked losing his license indefinitely. But Sergio was still jamming and kept right on going, thinking he had won the race through brute speed.

Limey Joe got his name from riding a Beezer, made in England. Although I had sold him my old Sportster that I had raked up myself, Limey's BSA was still his favorite, most dependable ride. The Sportster I sold him was a beater, a real bear to start. Limey had bruises up and down his legs from where the starter lever bounced back and hit him. Luckily, he lived on a hill, so it was usually easier to coast down the hillside and jump-start it that way. But when that wouldn't work, Linda and John would be stranded at the bottom of the grade, kicking and kicking until the flooded Sporty motor finally sputtered into action.

Limey was his own man. That's probably why he preferred the Beezer, despite the mountain of shit he took from the rest of the Club for not riding a Harley. Linda couldn't blame him, either. Besides, the BSA had a nicely curved sissy bar, the kind of bar that she could fall asleep on through the long distances. If she leaned back just right, she could balance herself perfectly and doze off for miles at a stretch. Linda logged many, many miles on the back of that rigid-frame bike, fast asleep, zonked out from the reds and the road fatigue.

Loaded Linda continued to live up to her name. She loved to get buzzed on reds, and whenever she dropped by my house at Golf Links to visit, somebody would invariably yell out, "Look! There's Linda, loaded again."

Pretty soon the name stuck. When she was hanging out with

Limey, she wore a T-shirt with a picture of a motorcycle and three L's scrawled over it, *LLL*—short for "Limey's Loaded Linda."

Linda's old high school buddy Deanna began hanging out, too. But unfortunately, Deanna started fooling around with heroin. Both girls would get pretty wasted. You'd see Deanna crashed out in the corner on a beanbag chair while Linda staggered around the room blindly, her hands out in front of her, hoping she wouldn't crash into anybody or anything. There was also a third gal pal, named Mary. She was really into smoking weed and would hit anybody up for extra reefer.

The three became inseparable around the Oakland Club, and one day Pee Wee called out to them, "Hey, look! Here comes Loaded Linda, Doper Deanna, and Marijuana Mary!"

Loaded Linda's appetite for Seconal soon put her in the red zone. She was getting too stoned for her own good. If it wasn't the reds that drove Loaded Linda, it was the Window Pane acid Tony laid on her for free. One night East Bay Jake had a party at his house on Seventh Avenue. Freewheelin' Freddy from Frisco (never one to miss a chance to drop free acid) was hanging out, too. Tony, the man with the plan, was crawling around the floor like a horse on all fours, with his old lady riding on top. Everyone was doing his or her own thing, drinking, smoking, eating, sleeping, fucking. Jake had the first Seeds album on the stereo, blasting a song called "Up in Her Room" over and over.

High on acid, Loaded Linda, Salem, and another chick named Little Lisa linked arms and bounced up and down on Jake's couch in time to the music. Jake started yelling at them to cool it; otherwise they were gonna break the springs in his couch. Jake's pad was no place to spill beer or puke on the carpet. He kept it up nice and expected everyone to respect his domain. Loaded Linda liked to hang out at Jake's, but some members were convinced that Jake's place was haunted. It all came about after a shooting that occurred in the house.

The so-called ghost at Jake's involved a girl named Jan who was

Pee Wee's old lady for about seven months. Pee Wee grew tired of her and was ready for a new girl toy. When he tried to break it off with her, Jan couldn't handle being dumped by a member. She became a stalker before the term was coined. She showed up everywhere Pee Wee was. If Pee Wee was hanging at somebody's house, Jan was knocking on the door, making a scene, or else calling him on the telephone.

It was "Pee Wee this, Pee Wee that." Everybody got really sick of it. One day Pee Wee was over at Jake's and Jan showed up at the house again. This time she ran upstairs to the bedroom, walked into the closet, found a gun, and shot herself in the head. Jan's brains were splashed all over the walls and ceiling. It was an especially bad scene for me because I was the guy who had to end up telling Jake that somebody splattered his closet walls with blood and guts. Boy, was Jake pissed off at Pee Wee when he found out.

After the shooting, weird things started to happen. The gun Jan used to kill herself that was kept downstairs kept reappearing upstairs. Somebody would put it back downstairs and the gun would be found upstairs again. Finally somebody buried the damned thing out in the backyard, and the gun *still* appeared upstairs.

According to Linda, the closet where Jan shot herself had a strange feeling. You could open the closet door and feel a cold draft inside. Trying to rid the room of Jan's nagging spirit, some members came over, tore the closet out of the house, and built a platform bed in its place. Even the toughest bike rider can't beat up a ghost.

Between the time when Elsie died and my second wife, Sharon, arrived, new chapters were forming in the U.S. beyond the California border. We were getting a lot of attention in the press—not just the local rags, but national magazines, too. I was swamped with business. At the time, Loaded Linda moved down the road from my place on Golf Links into Salem's house. Salem and Linda were just friends and she took care of

Salem's business for him, making his private runs and doing odd errands. After Elsie died, Linda always kept an extra eye on me to make sure I was okay. Her best friend, Patti, stayed at my house a lot and cleaned up around the place.

Because I was busy trying to keep guys out of jail, Linda and Patti would go out on the town and round up girls for me. It was like a scavenger hunt for pussy. They made a game out of going to strip bars and taverns, finding the cutest go-go dancers, and talking them into coming to our parties. After they'd bring one or two girls over to Golf Links to party, the girls would tell their friends. Pretty soon word spread to the point where every time Linda and Patti hit Telegraph Avenue in Berkeley on a Saturday night, girls on the street would recognize them immediately. Hippie chicks would walk up to Loaded Linda and ask, "Hey, where's the party? And how's Sonny doin'?"

Linda and Patti became party girl pimps. Patti could pick the prettiest. Linda would find the coolest. Whenever they'd disagree over which girl I'd probably like best, Patti usually won. Looks always beat out personality. If they brought a really good chick by the house, chances are I would keep her around for an extra week or two and let her hang out before passing her off to another member.

By the early 1970s, before I was sent to Folsom Prison, I had lost track of Loaded Linda. She got caught up in a beef involving counterfeit money. It was during the days of Richard Nixon and J. Edgar Hoover, and two Secret Service agents caught her red-handed passing bad bills and busted her. When I heard about what happened, I did what I could and made sure she was bailed out. But unfortunately, the Feds had a pretty open-and-shut case against her, and it looked like Linda was going to have to serve some hard prison time.

Linda did two years at Terminal Island before she was paroled. Her plan was to settle down with Limey Joe after she got out. But things changed when Limey Joe got into an accident on his bike on Highway 101 near Santa Barbara. He had busted his skull and died instantly. Pieces of his skull were littered across the pavement.

When Loaded Linda left Terminal Island, I slipped her fifteen hundred bucks to help her get back on her feet. I also knew the person who ran the halfway house where she was supposed to stay during her probation. He assured me she'd be treated okay. After that we lost touch again for many years, until we met recently at one of my book signings.

Linda has settled down and is doing really well these days. She has a high-tech gig in Silicon Valley and takes good care of herself. But every time a Harley-Davidson blasts by her on the freeway, the sound of the rumbling pipes takes her back to those early days when she was Loaded Linda, the notorious biker chick who rode on the back of a Beezer with her old man, Limey Joe—the night the pipes roared inside the Bay Bridge tunnel.

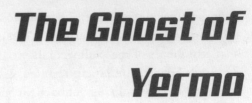

The Ghost of Yermo

When I'm riding long distances, I have my pipes and engine to keep me company. For me, it's not about being in a meditative state. I've had radios in my bikes for years, but I've never turned them on. I listen to my bike. I'm listening to what the belt, the exhaust, the chain, and the valves are saying. I want to know what the bike is doing. And it's talking to me. My advice? Listen to your bike when it talks to you.

Brenton had the Dago to Vegas ride down to a science. Four and a half hours, straight shot on Highway 15. In spite of his being a SoCal guy, the Strip was Brenton's favorite hang. There was always a hearty party in Las Vegas: strip clubs, peep shows, the casinos, fine food, and runs through the desert. Instead of flying or driving, Brenton preferred his 1993 Softail Custom ride.

Brenton hadn't been riding all that long, just about four years. Always fascinated by the biking lifestyle, he started out on a '77 FLH before the trade-up to the Softail. The guys at Moreland Choppers did most of his motor work, plus stretching and raking the front end. Adding a set of sixteen-inch ape hangers made for a stylish ride.

If Brenton chose to burn quick, Route 15 into Vegas served as perfect wind therapy. Like a lot of riders who cruised distances, Brenton used the time to reassess and speculate on life. How was he doing? Where was he going? What if he had taken a different path? Sometimes he dwelled on missed opportunities. These were the thoughts reserved exclusively for the long haul.

For a quick gas stop, Brenton usually pulled into the desert community of Barstow, basically a speed trap town right at the halfway point to Vegas. Next came a pit-stop town called Baker, home of the "world's biggest thermometer," where the temperature usually read somewhere between the high eighties and just over a hundred.

After an intense weekend of partying in Las Vegas, Brenton decided to dress light for the trip home. It was a hundred degrees outside. Since he didn't have a decent saddlebag, he shipped his clothes back to San Diego via UPS. All he carried was what was on his back: leather pants, T-shirt, and a denim vest. There would be no need for a jacket, since he hadn't planned on any late-night riding. The clock struck four as he passed the hot Vegas Strip, homeward bound. He'd be home before nine.

About ten miles past Baker, fifty-five miles out of Vegas, Highway 15 turned into a parking lot. After about a mile or two of lane splitting, the traffic intensified. Cars were overheated. People were hanging out in the median. Kids were running around. Car doors opened and closed. Brenton kept his lane-splitting pace safe and slow, in case he needed to hit the brakes quick. Then a trucker gave him the lowdown: a rig was burning out of control, and the freeway was completely closed down.

Brenton spent another twenty miles navigating the center lines, dodging more kids, animals, and obstacles. Once he passed the accident scene, it was smooth sailing again, full throttle down the highway. By the time Brenton hit the tiny town of Yermo, after being waved through the produce and dairy checkpoint, his bike began to sputter. He ignored the noise, but fifteen miles of lane splitting had burned more fuel than he thought.

By the next exit, he coasted off the freeway, stone outta petrol.

His bike died in the middle of a desolate exit intersection. There were no cars, no gas stations, no shops, and not even a phone. Just four corners of dirt. Brenton felt like a fool. Standing in the intersection with his hands atop of his head, he screamed and cursed his own stupidity.

Now he was stranded in Yermo, the capital of the middle of nowhere. Brenton had fucked up by not gassing up in Baker. Now he had to find a way out of this mess. He fought the urge to kick his bike over. A lot that would accomplish.

Then, out of the blue, Brenton heard a voice. He had no idea where it came from. He certainly hadn't seen anyone approach him in the wide-open desert.

"Son, you're out of gas," said the gentle voice.

The old man wasn't tall, maybe five-six. He had long white hair and a long white beard, sort of a cross between an emaciated Santa Claus and Father Time. With no shirt on, his belly was baggy and saggy; he was wearing only cutoff denim shorts. With no teeth and no shoes, the old man was a literal bag of bones, sipping from a bottle of Jack Daniel's, Brenton's drink of choice, though he offered him nary a swig.

"Is there a gas station nearby?" Brenton asked the man.

"Five miles down the highway."

Brenton didn't relish the idea of pushing the Softail that far.

The old man then told him about a couple who lived a half a mile or so off the highway. They might be able to help out, but Brenton would have to go down there by himself. Apparently the old guy didn't get along so well with them.

Brenton locked up his bike and started on his way. When he turned around to thank the old man, he was gone. Since there were no trees, bushes, or buildings around, the old man had disappeared as mysteriously as he appeared. Where could he have gone?

"Holy shit," Brenton thought.

Brenton was hungover. He'd partied pretty hard the night before. On top of that, he was pissed and tired. He wasn't much of a believer in the paranormal, but here he was stuck in the desert, talking to an apparition, which had simply disappeared. Strange.

Brenton headed down the road toward the house. He came across a large figure standing in a dusty yard. The guy was a huge dude, wearing overalls, six-six, big and fat. He was drinking God knows what from a two-gallon clay jug. His wife, missing her front teeth, sat on the porch. In the front yard was a pickup truck with what looked like someone's worldly possessions piled in the truck bed. It was the Yermo Hillbillies.

Brenton walked up and extended his hand in friendship.

"So, is there a gas station nearby? I ran out just up the road. You guys wouldn't have any spare gas you could sell to me?"

The man scratched his nuts and thought.

"I guess maybe I could siphon some out of the truck."

The man's wife (presumably) mumbled something about siphoning gas being illegal in California.

"Shut up," the man yelled to the woman. Then he turned to Brenton. "Okay, but you're going to have to siphon it yerself. I've got emphysema."

As the man brought the hose and gas can out of his truck, Brenton confessed that he had no experience operating an Arkansas credit card.

"Never mind," said the man. "I'll do it myself."

Sucking on the hose, the man coughed up a mouthful of gas.

Gas poured madly out of the hose, covering his bib overalls and spilling onto the ground. The man awkwardly navigated the gas-spurting tube into the gas can and quickly filled it. As he coughed and wheezed, Brenton tried hard not to laugh; the guy was doing him a favor.

Brenton offered the man twenty bucks for the gas, the can, and his trouble, and after much insistence, he accepted the bill. Just before walking back to his bike with the gas, he stopped in his tracks. Damn, he almost forgot; there was something he needed to ask the guy. Brenton turned to the huge man.

"Say, do you know anything about a little old man with long white hair and a beard? I ran into him up the road. In fact, he's the one who sent me down here."

The man and his wife looked at each other, puzzled, and shook their heads.

"White hair and a beard. Hmmm. Is that right?" remarked the man. "Was he kind of a soft-spoken guy?"

"Yeah, that was him," answered Brenton. "He had a bottle of Jack Daniel's in his hand."

"Sounds like my father-in-law, the old buzzard," said the man. "'Cept he's been dead five years."

At the next rest stop, Brenton was gassed and ready, a safe distance from the weirdness. Still ten or fifteen miles shy of the Barstow halfway point, the next challenge was to make up time and beat the cold. The closer he got to the Cajon Pass, the colder it would get. As he flew by Victorville (home of the late Roy Rogers and Dale Evans), the temperature dropped to fifty-eight degrees. It would surely drop down another ten degrees by the time he hit the Cajon Pass back toward San Diego.

The cold wind whipped straight through his vest and T-shirt. *Clack-clack-clack-clack.* Brenton was shivering so hard, he was sure he was going to chip his teeth. He thought about pulling over and stopping for the night, but he was determined to make it home. Stopping in Lake Elsinore, Brenton downed three cups of scalding

Brenton wailing down Interstate 15 at ninety miles per hour, breaking the helmet law: "I don't believe in ghosts, but I know what I saw."
(Photograph courtesy of Brenton Demko)

hot coffee, popped a few No-Doz, and was sufficiently cranked for the remaining seventy miles home. The bike seemed up for it. So Brenton kept the Softail on course until his frozen arms nearly gave out. Fifteen miles outside of Dago, freeway construction reduced the remaining journey down to one lane. It was half past midnight, eight and a half hours (nearly twice as long as normal), before Brenton finally made it home. After two hours of shivering under wool blankets, he eventually thawed himself out. Tomorrow for sure, he told himself, he would spring for some saddlebags and a thermos bottle. And the next time his bike talked to him, he would sure as hell listen.

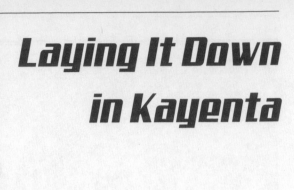

Laying It Down in Kayenta

Citizens tell me stories all the time about being broken down on the open road and how our Club guys will pull up and help out. Well, it's also cool when other bike riders turn around and help us out. And that's exactly what happened when Frenchie, a motorcycle rider from Mohave County, had the chance to be a Good Samaritan to the Cave Creek Club.

Frenchie rides a Suzuki Intruder and lives in northwestern Arizona. Mostly a lone rider, he loves scooting in the great Southwest, especially the ride from the Four Corners Run back through Kayenta, Tuba City, and other Navajo outposts, before heading back to his home in Kingman. It's a full day's open-throttle journey from Durango to Phoenix. The weather is usually hot and bone-dry, with only a few thunderstorms. The scenery on the isolated road is always spectacular. Outside of the high desert crosswinds that can whip in at over thirty miles an hour, it's a wide-open run through desolate Navajo reservation country.

Frenchie was on his way home from the 2000 Four Corners Run. It was Labor Day and the rally was over. Life was good. Frenchie had met up and partied with a bunch of his pals who rode in from Texas. That left a nice eleven-hour ride back to Kingman, a four-hundred-mile trek in all. Heading out on Highway 160, the road home was long and straight.

Because he had left at dawn, Frenchie was riding ahead of the usual post–Four Corners bike rush. It was early in the morning, with little traffic on the isolated highway. It was bright and sunny, with ultraclear visibility. A cloud formation hovered about fifteen miles down the highway. Red rock mesas dotted the horizon, and Monument Valley looked as picturesque as ever. Frenchie could almost see the jagged tops of Ship Rock miles away. Frenchie's Intruder hit its 80 mph sweet spot on the two-lane blacktop.

In defense of high speeds, the physics of a bike are like a gyroscope. Gyroscopes are more stable the faster they spin. Same with motorcycles: the faster you ride, the more stable you are as your senses kick into heightened alertness and you feel more "in the moment." If you hit gusty headwinds, sometimes it's safer to kick up into a higher speed, rather than slowing down. While such logic probably won't get you out of a speeding ticket, it may save you from a solo crash.

Frenchie was making amazing time on the 160. He had gassed up in Kayenta, and his next stop was Tuba City, ninety miles away, where he planned on grabbing some grub. As Frenchie putted along, he looked into his rearview mirror and saw two columns of headlights roaring out of Kayenta. He knew full well what was coming up behind him, twenty Cave Creek Club riders hauling butt, two by two in extratight formation, like fighter jets. Frenchie was still doing eighty, but the pack was gaining on him fast, pinning

100 mph. Our pack passed him as if he was standing still. As Frenchie let out a holler, he gave them a clenched-fist salute.

About a quarter of a mile further down the road, a Navajo cop in his Bronco/Blazer (whatever) was stopped cold in the middle of the lane. Maybe the cop was thinking about turning left, or maybe he had his thumb stuck up his ass, but the pack blew around him at warp speed. We didn't even bother to slow down. What the fuck was he going to do? Enter a high-speed chase with twenty screaming 1%ers on Harleys?

Thirty seconds later, the pack was down the road, clean out of sight. As Frenchie throttled down, another Harley came barreling up from behind. He barely saw the bike in his rearview mirror as it came up from behind pretty quick. It was another one of our guys. The straggler was Marco, packing his old lady, hoping to catch up fast with the pack. He was riding a full dresser with crash bars on the front and smaller ones on the back for his girl. As Marco hit optimum speed, he passed the Intruder and was positioned to pass the cop's truck on the left. But there was a car coming head-on in the left lane, so Marco pulled quickly back into the right lane ahead of Frenchie.

Then everything went into slow motion. Frenchie could see Marco speeding toward the parked truck. No way was Marco going to be able to slow down and stop in time. As soon as he realized what was going to happen, Frenchie pulled over and, with very little time to downshift, skidded to a stop on the dirt next to the highway. Meanwhile, Marco locked up both his brakes; it was obvious he was going to crash into the cop in the Bronco. He was going to have to lay his bike over, but not at 90 mph. Marco kept the Harley up on two wheels as long as he could, jackknifing back and forth. From the roadside, Frenchie knew right away where Marco was going with this maneuver. He would jam it up and lay it down at the absolute last possible moment. As Frenchie watched it all go down, he sensed that Marco knew exactly what he was doing. At that fateful moment, Marco slowed the bike down to twenty-five, not too bad a speed to

drop the bike down, considering how fast he had been going.

Like a pair of daredevil stunt riders, Marco and his old lady slid underneath the truck. The dazed Navajo cop climbed out of his truck, but Frenchie was the first at the scene of the accident. Marco and his old lady were both already on their feet. His saddlebags burst open from the impact, his effects scattered all over the road. The two seemed more stunned than injured. Marco's old lady was walking around with only a few abrasions. Frenchie passed her his water bottle, asking her if she was okay. Considering that she'd almost been killed, she shook her head with uncertainty.

Other bikes began pulling up as the traffic started to pile up. After he helped Marco roll the bike to the roadside, Frenchie decided to haul ass to Tuba City, sixty miles down the road, and hook up with the rest of our pack to give us the news.

Frenchie revved up and took off like a shot, making the ride in less than thirty minutes. He spotted our bikes parked outside a coffee shop and gave us the lowdown. The pack sent a prospect with a truck back to pick up Marco, his old lady, and the busted-up bike. After witnessing the best bit of motorcycle crisis control he had ever seen, Frenchie was back on the road to Kingman, having done his good deed for the day.

I guess there's a lesson here. When you're riding hard, always have an escape route in mind for when the inevitable road hazard occurs. Any good motorcycle rider has an innate set of what-if contingencies should something unexpected occur on the open road.

Marco knew exactly what he was doing at that moment. He was riding the bike on sheer instinct, keeping it up on two wheels for as long as he could keep control and slow down. Had he ridden into that truck on two wheels at full impact, Marco and his old lady would have been smoke.

Apple-Pickin' Time in Sebastopol

There are situations that happen to everyone, and at the time they may seem like a disastrous turn of fate. But revisited after a period, and seen from a new perspective, even the worst experiences seem more humorous than distressing.

This story is one of those.

In late spring of 1965, Sebastopol was considered the "Apple Capital of the World." Everyone in this small Northern California community depended to some degree upon the local harvest. The hills that were covered with a carpet of white blossoms had given way to rows and rows of green orchards. With the exception of light seasonal rains, the weather was warm and perfect.

This was the first baseball season Randell had missed since entering Little League nine straight seasons ago. But baseball wouldn't figure into his future. He was leaving for Marine Corps boot camp right after the final apple picking.

It was natural for Randell to spend all day Saturday, one of his last days in town, on a high ladder thinning apples. Maybe this process should be explained: Apples grow naturally in clumps of four or five. In order for one to mature to a marketable size, the best of the bunch is selected, the rest discarded. This is very hard labor, but at ten dollars for a ten-hour day in 1965, it seemed worth it.

This Saturday was extraspecial. Randell had gotten a date with a girl we'll call Jane, who was considered by most to be the prettiest girl in neighboring Santa Rosa, a California town proclaimed by famed horticulturist Luther Burbank as the "City Designed for Living." Randell's mood was cheerful as he worked through the day without breaking for lunch. He needed to leave a little early so he could scrape the dirt off of his old military surplus Flathead 45, the bike he had bought through an ad in the *Press Democrat* for fifty bucks.

It was near the end of the thinning process, and all of the Gravenstein apples looked good enough to eat. So Randell snacked throughout the day on the freshest green apples in town. This late in the season you could barely notice the slight tartness as Randell bit into the shiny green and golden skins.

While Randell wasn't exactly nervous, the thought of dating a rich girl was a little daunting. Randell never expected that a babe like Jane would consider going out with him, a mere biker peasant. If Randell hadn't stopped at the Baskin-Robbins 31 Flavors while "tooling" 4th Street the previous Saturday night, he might have gone his entire life without knowing this girl even existed.

Everything about her intimidated Randell from the moment she marched up to him, in front of all his friends, that Saturday. Randell and the gang had been hanging around the parking lot of the ice cream parlor when Jane boldly stepped right up in his face, introduced herself, and slipped Randell her phone number. This chick took control in a way that Randell had never experienced. When, the following evening, *she* asked *him* out, Randell was just barely able to mutter a stunned reply.

"Sure, why not?"

The night of the big date, Randell could see his face in the twenty-nine coats of black lacquer paint on his tank, while the chrome reflected his pegged Levi's and Bates Floaters. Randell cruised down Fourth Street looking for his buddies, the very guys who would recharge his confidence. Instead, his older sister Sandy stopped Randell on the street. She had escaped the farm a couple of years before to attend a local college.

When Randell was young, Sandy was like a mom to him. Their mother had it tough, widowed and stricken by polio in the early 1950s, in and out of one hospital or another. Sandy tried to fill in for his mother as best she could. It was a debt Randell would always remember.

Sandy was excited, but she needed a favor. She and her girl-friend had just rented a new apartment in Santa Rosa, so could Randell help her move a mattress and box spring up some stairs? It was a nice little two-bedroom apartment in a safe middle-class res-idential area, and other than the one bed and a few boxes, it was completely empty. Come Monday, the water and electricity would be turned on so Sandy and her roommate could move in.

Sandy persuaded a reluctant Randell. He still had plenty of time, but man, he didn't want to mess up his clean Levi's or his new madras shirt. Sandy gave Randell the keys to her empty apartment anyway, and he moved the mattress in record time on his way to Jane's.

Jane and her family lived in the middle of guarded luxury in a neighborhood of the wealthy elite. Jane was from a very different world and, with her rich family, moved in all of the "right" circles. Randell had never known anyone like her. He had assumed she dated exclusively within the country club set. Jane's clique only considered talking to guys like Randell during football season, as if somehow the game transformed them, if only for a few months, into their elite power group. What was she doing with a guy like him?

Now here he was, a tall, lanky boy, standing awkwardly in Jane's living room being scrutinized by her parents. Luckily her father was a football fan. Instead of wanting to know why he rode a god-damned motorcycle, he was curious about Randell's plans for the following season. Randell explained that he was too green for the full-ride scholarship he was offered by Cal-Berkeley (the old man's alma mater), and besides, Randell would be going into the Marine Corps to play football for *them*. Randell could see the blood drain from Daddy's face as he retreated quickly to refresh his drink. Mercifully, Jane bounced into the room, grabbing Randell's arm and waltzing him out the front door.

The expected "Jane, you be careful on that thing" did come, followed by the "Shouldn't you kids be wearing a helmet or something?" Jane wore a tartan skirt and was braless under her angora sweater. Off the steep driveway, a coasting start was easier than kicking, so Randell and Jane accomplished their quick escape. Packing Jane was a dream, as she was a quick learner; within minutes she shifted her weight and leaned like an experienced biker chick. Every time they came to a stop, the fumes from Randell's perpetual gas and oil leaks were replaced by the scent of her expensive perfume.

A giant bug-magnet smile grew across Randell's kisser, and as the two leaned into a hard downhill switchback, Randell grabbed third as they came out the backside. Jane was stimulated by the energy and power of the bike between her legs. As if she were on a roller coaster, the faster and more perilous her first ride seemed, the tighter she squeezed. And Randell couldn't stop biting his lip; those were Jane's arms, and she was digging his world. Randell slowed as the road flattened out, and Jane bit the bottom lobe of his ear.

"Jane, what would you like to do? There's a good movie at the California Theater."

Thinking for a moment, Jane leaned over further and kissed Randell very affectionately on the cheek while rubbing his inner thigh.

"No, Randell, we don't want to go to any old movie. I want you all to myself!"

Randell nearly curbed the bike after that response.

"Why Jane, what do you have in mind?"

"Do you have anything to drink?"

In 1965, no self-respecting guy would think of going out on a Saturday night without toting at least a couple of six-packs.

"I have a few beers in my saddlebag. Coors or Colt 45?"

"I'll have a Colt. Then why don't we ride up toward Sonoma and park?"

Randell was likin' what he was hearin'. Beer and a blanket, life could be so beautiful.

Most of his high school buds assumed Randell was a ladies' man, but truth be known, Randell was shy and not very experienced with the girls at all. Randell was still thinking in terms of bases— first base, second base, third base, home plate—but he'd only been able to get a good swing at the ball a couple of times, so to speak. And if there's anything on this planet as horny as a seventeen-year-old girl, it's an eighteen-year-old on his way to the Marines. Jane's aggressive behavior was certainly way beyond anything Randell had ever experienced.

The beer was warm, the sunset was photogenic, and the necking was sloppy—what a great moment to freeze in time. The only thing that was not quite perfect was that Randell's stomach seemed to growl every time he moved. Within an hour, the petting escalated into heavy groping. It was time to make the big move. Coming up for air, Randell suggested they head out to his sister's new apartment.

"This is Jane, beautiful Jane!" he thought. "And she WANTS ME!!"

Randell, the poor biker farm boy, was poised to score a goal, big-time.

Randell wasn't sure he'd even make it to Sandy's apartment, the way Jane was sucking hickeys onto his neck and continually rubbing his crotch. Racing as fast as he could, he screeched the bike to a halt on the sidewalk next to the stairs to his sister's empty apartment.

The thought of no lights or water didn't concern either of them. They hadn't even bothered to spread their blanket over the bare mattress. Clothes dropped wherever they fell on the bed as Randell tried everything he knew twice, and then a few things he'd only heard about or seen in a couple of James Bond flicks. For about two hours, Jane and Randell escaped into sexual bliss.

It was getting late, about 1:30 in the morn, as Randell lay back spread-eagled, folding his hands behind his head. Feeling like Caesar the mighty conqueror, he wore the smuggest of grins across his face. Jane's skin was soft, and her nude body was beautiful in the moonlight. Tweaking her nipple, Randell remarked, "You little softie."

Feeling playful, Jane punched his stomach.

Very big mistake.

In a reflex move, Randell tightened his abs as his intestines exploded and diarrhea sprayed, covering Jane and the whole bed. The mixture of green apples, warm beer, and hot sex had done the trick, and the results were nauseating and sickening.

"Oh, shit!" Randell screamed, springing to his knees. Jane gasped, appalled at the unexpected biological flood. She heaved, then coughed, vomiting on Randell's chest. At the sight of so much errant body fluids, Randell then hurled everything *he'd* eaten for a week all over *her*.

Without the apartment's water being turned on, it was impossible to clean up. Randell's and Jane's clothes lay beneath them on the mattress, so neither had any other clothes to put on. Fortunately, Randell did have a pair of gym shorts in his saddlebag, and the blanket wasn't *that* bad. But Randell still needed to get Jane cleaned up before she could go home.

Not far down the street, a late-night supermarket's and a gas station's lights were still on. The two tiptoed out the door and jumped on the bike. Not wanting to alert anyone, Randell cut his engine and coasted into the back of the filling station, where they would then bolt for the rest rooms.

Of course both restrooms were locked.

Jane was nude, wrapped in a blanket, standing in front of the locked door to the ladies' room. Her whimpers soon turned into uncontrollable sobs. Randell never felt so bad for anyone as he did for her at that moment. Then, as if things couldn't get worse, both the gas station attendant and a carload of Jane's girlfriends appeared from around the corner. There Randell and Jane stood in front of Randell's motorcycle, nude and soiled. The response from the attendant and Jane's girlfriends was a mixture of horror and hysterical laughter. As Jane's girlfriends whisked her away, she probably never heard Randell's apology. A couple of days later, Jane left town and went away to college. Randell was on his way to Vietnam.

They would never see each other again.

Randell learned two important lessons that night, lessons in life he would never likely forget: First, never eat too many green apples. Second, Shit Happens!

On the Lam

In June 1979, the Feds launched a raid against us pursuant to the RICO statutes. Racketeer Influenced and Corrupt Organizations: Fancy words for a lot of bullshit. The Feds said our motorcycle club was operating as a group criminal conspiracy. They hit a bunch of our houses, our friends' houses, our clubhouse, and about every other place they thought we might be, trying to round us up. A few of us got away and became fugitives.

Cincinnati was among this group. He had an interest in some gold mines in a tiny town in California's Gold Country. This town was located at the end of a nineteen-mile dead-end road, which made it pretty easy to keep track of any comings and goings.

Late one night, No Name, another RICO

fugitive, showed up in town to visit, relax, and lie low. Cincinnati was happier than hell to see him. Just before daylight, they took off in Cin's van to do a little sight-seeing. Having spent many days in the area with nothing to do but explore, there weren't many places Cincinnati hadn't been, so he was a more than competent guide.

They drove about twenty miles further into the woods and ended up on a river gorge where the guys spent most of the day exploring several old mining claims. By the middle of the afternoon they headed back toward town. On the way back, they noticed vans with men in them carrying weapons. The two RICO fugitives took a quick detour to the upper part of town, since Cincinnati knew the people who lived in the very first house off the road.

The lady who lived there came running out to tell them that every cop in the area (and then some) was in town, looking for Cincinnati. They ran into her house and went upstairs to the bedroom so they could check the lay of the land and find out what the fuck was going on. Cincinnati was peering around with a pair of binoculars. As he was looking at a trailer home parked down the road, he spotted someone peeking back at him through binoculars. Talk about an adrenaline rush.

"Who lives in that trailer?" Cincinnati asked the lady.

"Hillside Hal. He's pretty harmless."

"What time do you usually wake up?"

"About this time."

"Where do you get dressed?"

"Right where you're standing."

False alarm. Man, it wasn't anything but a peeper taking in some morning viewing. Cincinnati told her not to disappoint her fan club, which she didn't.

Cincinnati and No Name jumped back into the van and made a fast U-turn. Instead of going back up the hill, they took a lower road. The road forked. They went left, since the right fork led to an old mine that the town used for its water source. After about two miles, the two stashed the van and ran down the hill, packing Mac-11s with four or five extra clips.

At the bottom of the hill was a dry streambed. After they made it down there, Cincinnati and No Name followed its course. Vegetation grew over the streambed, forming a tunnel. They followed the tunnel until they reached a wall of rock that must have been one hell of a waterfall when the water was flowing. Stopping to catch their breaths, since it was getting close to nightfall, they decided to wait until full dark before moving on. But it would be a short rest, as they heard sounds in the brush. Was it someone from the posse? As the noise got closer and closer, the two moved deeper into the shadows, their guns raised and ready.

Then, rounding the corner, it appeared: a five-pound skunk had panicked two big bad desperados. They let it walk by without moving a muscle. It seemed like a good plan to keep moving, so the two crawled up the rockface and came to a small plateau. Sitting there was a town.

"What's this place?" No Name asked.

"I've got no idea in the fucking world," said Cincinnati.

"I thought you knew this area like the back of your hand."

"I sure as shit thought I did, but this is either a mirage, or we've slipped into a parallel dimension. I ain't never seen a place like this."

The little town consisted of a small row of neat and trim houses lining each side of a dirt street. Each house had wrought-iron fences, walkways, and porches. They were probably built in the late nineteenth century; they were most definitely unoccupied and had been for a long while. Walking down the middle of the street, they followed the curve to the right. The buildings and the mine tunnel of the old mine that supplied the area's water suddenly appeared.

"Shit," said Cincinnati, "I know where we are now. We just came in from the back end. Man, this place is cool. I wish I had checked it out sooner."

Cincinnati had been in the area numerous times but had never had a reason to walk this far around the corner. From where they were standing, they could see the town and a streetlight. Away from its beam was total darkness, and beyond lay another old mine warehouse, a great hideout. But there was a large opening at the far end that would have been a perfect place for an ambush by the Feds. A very large door once closed the entrance, but it had been removed many years ago. Past the warehouse, the darkness would make it very hard to be seen from town.

Cincinnati handed everything to No Name and told him that if anything happened, he should run like hell. He took a deep breath, gritted his teeth, and moved out. He did not like the idea of being there. Cincinnati was convinced that at any second, his worst nightmares were going to come true. When he passed through the warehouse and reached the loving presence of darkness, he breathed a large sigh of relief and turned to wave No Name forward.

When he turned, his pal was already right behind him and damn near ran full-face into him. It startled Cincinnati so much, he damned near shit. After a few more adventures, like untangling No Name's hair from the tree branches, the two came full circle and reached the house where they had been warned earlier that day.

The van that they had driven had been retrieved and was parked just up the road, with a perfect view over the whole town. The lady of the house (the one who had warned them earlier) was still there. She had more news. Fifty or sixty cops had hit the town and had sealed off all the exits. They even went so far as to stop the school bus and search it.

Earlier that day, Cincinnati's lady had managed to escape. Some kids in town had arranged to hide her in a small cabin in the dead center of town right under the Feds' noses. That day, they had stuck wanted posters up all over town, offering a $10,000 reward for any information leading to Cincinnati's capture. As fast as they were sticking them up, the kids were tearing them down. And while the cops were going crazy chasing the kids, another group of kids led Cincinnati's lady to her hideout. She was waiting for word before making any kind of move. After a few hours, most of the cops had left, except for one deputy sheriff hanging tight in town, just in case. Cincinnati made his way to the cabin where his lady lay in wait.

At the house next to the cabin, there was a wedding party going on that evening. Things got wild and the groom fell and cut the hell out of his hand. Another local friend of Cincinnati's, the captain of the volunteer fire department and also the town ambulance driver, was at the wedding. The deputy, who was also hanging out at the wedding, didn't think the wound was severe enough to warrant an ambulance trip into Nevada City, thirty-five miles down a winding road. Cincinnati's friend disagreed, and reminded the deputy that since he was the only authority figure present, it was his responsibility to take care of things.

Cincinnati's friend went for the ambulance after he dispatched several kids to Cincinnati and No Name's hideout with the message that they should follow as soon as they saw an ambulance head down the hill, since all of the law would be inside.

As soon as the ambulance was loaded with the bride and groom, it headed out. About half a mile behind was a car carrying Cincinnati, and behind him was No Name in his van. As the ambulance headed west toward the Nevada City hospital, Cincinnati and his friend went east to safety, to spend more time on the lam.

Take the Long Way Home

A **lot of women bike riders may be** better riders than men because their reflexes are quicker. But it gets down to the individual and her experience. When shit happens, normally their reactions are faster. That's why females are often such good dragster riders. Off the line, their reflexes can be better than a man's. I once had an old lady who used to drag-race and she would beat the guys off the line so bad it startled them. Even if they had faster bikes, she'd still beat them. She rode an 89-inch Sportster and pretty soon the fuckin' guys wouldn't race her. They'd think, "If I beat her, I beat a girl. If she beats me, I got beat by a girl." She was in a lose/lose situation. If her riding skills sucked, she'd get shit; if she rode like a champ, men would chill her out. Who wanted to get shut down by a girl on a Harley?

Debbie was an exceptionally tall woman with long legs. Her bike-riding boyfriend, Jackson, was built like a pro football line-backer. Jackson, a potential 1%er, rode a '69 Shovelhead with a mustang tank and a suicide clutch. You probably wouldn't call Debbie and Jackson a perfect riding fit. Because Debbie was tall and long-legged and Jackson was bulky and big, the two rarely enjoyed packing together on long distances. Like a lot of chicks on bikes, Debbie soon got tired of riding backburner. She needed a motorcycle of her own.

Debbie knew that if she was ever going to pursue her dream of riding motorcycles, it would have to be on her time, on her dime. She was going to have to learn to ride by herself, probably on the sly. Debbie waited for just the right time to ask Jackson to let her take his bike for a quick spin around the block. Finally one night, Jackson fucked up and did Debbie so wrong (another girl, as usual), she had a marker on her man. When she asked to ride the bike, Jackson could only grunt and throw the keys to the Shovel her way.

Debbie rode around the block a couple of times without a spill. Then she called her best pal, Jeanette. Maybe they could ride around the neighborhood together? Jeanette's boyfriend was much more supportive of his old lady riding motorcycles than Jackson would ever be. So much so, he built Jeanette an awesome orange three-wheeler with jumbo full-sized tires and custom chrome rims. Made from an old Panhead, this trike was a hog and a half. Jeanette, also a babe, turned lots of heads when she rode it around town. Jeanette and the three-wheeled monster were a regular attraction at the Oak-land Roadster Show. The trike was dangerous for anybody to ride, even a skilled rider like Jeanette. They frequently tipped over. When she rode with other bikers, the trike was always—without excep-tion—trailing the pack. Jeanette was careful to make sure the tire pressure was as low as possible, to keep the darned thing from bouncing at the slightest pothole or road alligator.

Jeanette and Debbie took off together for one more scoot before the sun went down. Jeanette was jazzed to be riding with another chick. As they took off around the corner, neither wanted to ride only a few blocks and back. They were going to bolt and take the "long way back home."

The "long way back home" was a rural stretch called Kirker Pass Road, a wooded two-lane road a few miles from where Debbie and Jackson lived. As Jeanette and Debbie putted along, they came to the long, downhill hairpin section of the road that bottomed out into a ninety-degree turn. Every day in her car, Debbie took that steep part of Kirker Pass Road for granted. But this time she was riding her boyfriend's prized Shovel, and for a first-time bike rider, Kirker Pass seemed about as steep as a roller-coaster ride and as ominous as the Grand Canyon.

What the heck. Debbie let out a long, deep breath and started her descent. She gained more and more speed as the sharp turn at the bottom approached. It was too late to turn back, and Debbie freaked out. Her arms shook nervously. Her hands gripped the handlebars like vises. To make matters worse, the bike's motor abruptly cut out as she continued her downhill drop. Debbie wasn't going to make it. With absolutely no throttle control, a stalled engine, and the bike picking up more and more speed, Debbie couldn't control the Shovel's velocity. Jeanette, immediately sensing her friend's danger, held back the traffic on Kirker Pass by weaving broadly back and forth across both lanes, Highway Patrol–style. Debbie's Shovel was still gaining speed, and Jeanette saw the bike lean sideways. Debbie could feel her leg grazing the road's surface. Her heart was pounding. All she could hope for was that when she laid this bike down, it would slide in such a way that she wouldn't be killed.

But Debbie was no dummy, and she had good instincts. At the right moment, near the bottom of the hill, she pulled the clutch in one last time, shifted, and released. By the grace of the gods, the motor kicked back in. The force and the inertia of the engine pulled

Debbie on her boyfriend's '69 Shovelhead minutes before her perilous *downhill cruise.* (Photograph courtesy of Debbie Tolly)

the bike back and slowed its velocity. She was no longer in free-fall mode. The engine's rapid adjustment of speed literally jerked the bike back upright, and Debbie managed the sharp turn at the bottom of the hill with a few feet to spare.

Debbie and Jeanette pulled over to the side of the road. Debbie's hands were now drenched and shaky, her panting nearly turned to tears. There was a nervous pause as Debbie and Jeanette composed themselves. Then, almost simultaneously, the two women let out a loud, gleeful scream.

"Fuckin'-a, Bubba! Let's do it again!"

Debbie and Jeanette kick-started their bikes and headed back home. Jackson was never going to hear about this.

I t wasn't long after her near spill on Kirker Pass that the time was right for Debbie to give Jackson the boot. She was tired of his antics and, more important, she finally had enough money in the coffee can to buy herself a slick Harley Super Glide. The Super Glide was the perfect bike for Debbie; she liked riding low on the road and felt more in control with a longer scoot. After a few months with her Super Glide, she gained enough confidence to join her guy friends on an extended Northern California motorcycle run to Reno and Lake Tahoe.

Riding with a group helped Debbie become a better rider. In the earlier days of riding "rigids," as skillful and competent a rider as Debbie was, she could never quite match the speed and endurance of her male biker friends. It was a challenge keeping up over the long haul. Navigating difficult weather conditions like rainstorms and cold, windy days was an even bigger challenge. But for all that, the guys liked having Debbie along. She could hang out with the best of the guys and not care about being the only girl in the pack.

That weekend Debbie showed up for the huge sponsored summer run with about forty other bike riders. Debbie wore her sexiest leather halter top, the standard uniform of the day. This was the biggest group she had ever ridden with, and as she took her place in the middle of the pack, for the first time Debbie experienced the formidable and irrepressible power of forty pairs of Harley pipes trumpeting at her in unison. The moment was overwhelming, and Debbie felt magnificent as the riders kicked into high gear together, unleashing a tidal wave blast of unbridled torque.

Everything was going great until a wasp flew inside her leather top. Debbie held her position as best she could while trying to shift her halter top to get the wasp out. But the more she tried to readjust her top, the more the wasp stung her, and the little bastard just wouldn't buzz off. As a last resort, Debbie pulled the rip cord, that is, the string on her back. With the wind whipping at high speed, the leather top flew off and into the face of the guy behind her. To the delight of the fellows who surrounded her in the pack, Debbie rode topless for the next ten miles before the whole pack could find the right place to slow down and let her pull over. As Debbie rifled through her saddlebag for a support T-shirt, one of the guys managed to grab his camera for a Kodak moment. Who could blame him? The pain was almost as severe as the embarrassment, and Debbie learned that it was worse to have a bee in her halter top than it was to have one in her bonnet.

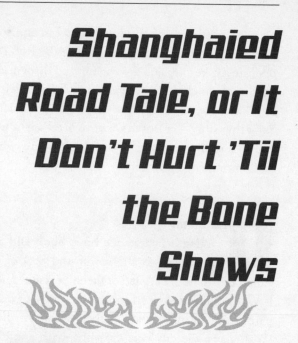

Shanghaied Road Tale, or It Don't Hurt 'Til the Bone Shows

Almost every year, I get an invitation to ride in Japan. Although we don't have a Club chapter there, the American biker lifestyle fascinates the Japanese. A run in Japan is very different from those in other places in the world because the country is relatively small and crowded. So when Japanese motorcyclists go out on a run, they stop every fifteen miles, have a cigarette and a bowl of soup, and talk. The idea is to make a hundred-mile run last all day.

I've never ridden in the People's Republic of China, but I did get some correspondence from Butch, an Australian bike rider based in Shanghai. Butch gives a clear picture of what it's like to ride a Harley in China and, worse, what happens when you crack up on one. Butch starts off with a little background on Shanghai and some commentary on being a rider there. He and his crew, the Red Devils MC Shanghai, operate a bike club and a web site that dispenses biker information, such as the nightmare of licensing a bike or getting a driver's license in China. He describes his personal, hell-raising motorcycle tale. Take this story—in Butch's own words—as a warning, especially if you're seriously thinking about doing some riding in a Third World or Communist country.

Dear Sonny:

What a history we have here! Old Shanghai from 1920 until 1949 was the worst and best of everything. It's been called the "Whore of Asia," "Queen of Asia," and "Paris of the East." It was a paradise for adventurers, a city of quick riches, ill-gotten gains, and fortunes lost on the tumble of dice. During its "Wild West" days, the city was crawling with gamblers, swindlers, drug runners, idle rich, dandies, tycoons, missionaries, gangsters, and back-street pimps.

The city had such a bad reputation in certain quarters that it gave rise to the popular saying "to be shanghaied," which meant "to be drugged and shipped off to sea as a sailor." Historically, the saying refers to the problem ship captains often had when they arrived in Shanghai and needed to put together another full crew to sail off again. Nowadays we're still getting shanghaied, but in a very different way.

Every other night of the week, while sitting in my favorite watering hole—O'Malley's Irish Pub in downtown Shanghai, China—there is always some foreigner butthead biker-wanna-be who comes over and bugs me while I'm enjoying my favorite beverage, ice-cold Victoria Bitter. It never fails.

They ask, "Do you ride or just wear the clothes?"

After my reply of "Hell, yes, I ride!" it always leaves them wondering. Then I get to hear about their Hardtails, Pans, Shovels, and, of course, their new Arlen Ness bikes. These poser fuckheads own every bolt-on-chrome and plastic part they can get on their credit cards, but a bike? Where is it?

I love to see their faces when I ask them where their bikes are now. That's when I get to hear some of the poorest excuses ever uttered:

"My wife made me sell it."

"The company wouldn't let me bring it."

"I heard the gas here was no good."

"I was told there was trouble getting parts."

"Insurance is too high."

And of course, my very favorite: "I left it in storage."

Holy shit, most of these geezers are only here for one or two years!

Then, when I tell them I have a Sportster and a Dyna in Shanghai and a Softail in Beijing, they want to give me shit about the Sportster! I tell them, there's a lot of history behind the Sporty, but fuck no, these puke-faced motherfuckers just want to talk shit about how cool it is to ride full-sized bikes and have the wind blowing in their hair. Shit, these are all clean-cut suit-and-tie guys, and there isn't any hair to get in the wind to begin with! Don't they realize that the word "bro" or "mate" gets worn out if everybody is a fuckin' "bro" or "mate"? These assholes then proceed to get shitfaced and embarrass everyone who ever rode a motorcycle.

When I go to leave and sit on my Sportster, all I ever see parked around me are mopeds and some shit 125 cc bikes that I never heard of in my native Oz. I really get a shit-eating grin as I fire up those eighty-nine cubic inches and rev the mill through the drags. Local men stick their fingers in their ears, and women grab their kids and duck into shops. I set off car alarms everywhere.

I wear Harley shirts, but I ride, too. I ride no matter what the weather or how much the insurance is. There's always octane booster for the shitty gas. As for the wife, she would never come between me and my bikes (my Harleys and Chang-Jiang with sidecar).

 **Butch hitting the streets of Shanghai, setting off car alarms and terrifying the natives.** *(Photograph courtesy of Butch Walter)*

A couple of years ago I survived a very nasty accident. I almost lost my right arm riding the first Harley Sportster in Shanghai. Finally, after 12 operations and 105 hours total operating time, I'm on the road to recovery. No, I wasn't drunk or stoned, either; I just got run down by a stupid farmer from some hinterland province making a living as a taxi driver in the big city of Shanghai. Without the assistance of my very good Kiwi mate Peter, his son, Shane, and my lovely wife, Wendy, I would have lost my right arm or even died. The first emergency operation was done in a local Chinese hospital. The moment my mates arrived at the emergency room, the doctors stopped trying to amputate my right arm with a hacksaw. My wife was in a different section of the hospital sorting out the payment issue before they would start the proper surgery.

The hospitals and doctors don't recognize any foreign insurance in mainland China and they always demand cash (U.S. dollars or Chinese renminbi) before any treatment. During the first night and operation, my mates had to fork out the cash left and right, and even pay taxi bills for some of the doctors and nurses who arrived on the scene. The Chinese doctors tried to operate for fifteen hours, with everything going wrong that could go wrong, including giving me two liters of the wrong type of blood. Afterward, they put me up in a ward with twelve Chinese workers injured on a construction site. I had to go through hell without the proper medication to kill the pain. Eventually I had blisters the size of golf balls on my heels from rubbing my feet on the bedsheets due to the tremendous pain.

My wife and mates were finally able to contact my insurance company in Hong Kong. The company said it would fly a doctor up immediately, someone who would evacuate me to Hong Kong for further treatment. Due to a typhoon in Hong Kong, the airport was closed. I was still stuck in the local hospital, where they took my temperature with those damned electrical thermometers used to check air conditioners. Hundreds of Chinese spectators passed through, since it wasn't every day a half-naked, huge, hairy foreigner was on display behind a glass wall in one of their hospital beds. Finally my doctor arrived from Hong Kong on Sunday afternoon, almost forty-eight hours after the accident, and saw all the mess inflicted on my arm by the local doctors in the hospital. Without any strong painkillers, I was really relieved to see a real doctor.

The transfer by plane down to Hong Kong was a blur to me. My doctor injected liquid painkillers into my spine, enough to put a horse down. Once I got to Hong Kong, the doctors, nurses, and insurance company took excellent care of me in Hong Kong's Queen Elizabeth Hospital. I was released and sent home with half of my forearm missing and extensive, severe damage to my nervous system. It took another ten weeks to rebuild my forearm with spare parts from both of my legs, including muscle, nerve, and skin grafts.

After my release from Queen Elizabeth Hospital, I had follow-up work done on my wrist at the most sophisticated hospital in China, HuaShan Hospital Shanghai, the late Deng Xiao-ping's hospital as well. An outstanding team of nerve and hand surgeons (recommended by the same Hong Kong doctors who saved my arm) worked on me.

During the rebuilding process of my right forearm, I started my own therapy, to seriously rebuild my own damaged Sportster. With the help of my mates in different countries around the world, every single replacement piece for the Sportster had to be hand-carried (smuggled) into Shanghai.

Butch's reconditioned 89-cubic-inch Sportster, the bike that nearly cost him an arm—literally. (Photograph courtesy of Butch Walter)

Since my cast came off, I'm now able to hold my dick with my right hand for a piss again and, yes, take my Chinese Chang-Jiang with sidecar (a Chinese BMW R71 copy) out for a 220k ride through the Shanghai countryside. It was eleven months before they let me ride again, and I'm just now recapturing the spirit. By the way, after you get accustomed to the feeling of half a leg attached to your forearm and the hair starts growing again, it's not that bad after all. Now I just have to design a tattoo to cover up all the scars.

Believe me, the Sporty ride is coming along, hopefully done in time before the next anniversary of my accident, louder, faster, repainted, and coming to a bar near you soon. The Dyna and Softail are ready to rumble, too, but as a statement of proof to myself, I will drive the rebuilt Sporty 89 cui for some serious tire burning and wheelies around downtown Shanghai.

Total breakdown of damage:

- $360,000 (U.S.) in medical bills, fully covered by insurance.
- Approximately 3,800 stitches to my right arm and both legs (including muscle and skin grafting).
- Sixteen total weeks in the hospital.

My Harley and I are both now genuine rebuilds, not yet perfect, but running very strong. The moral of the story: Make sure you have a very good health insurance policy when you ride in China, and also some mates who will be there when you need them.

Sincerely,
Butch
Red Devils MC Shanghai

Experts, Judges, and Covered Wagons

've been associated with 1%er motor-
cycle clubs since the late 1950s, and believe me, motorcycle clubs are exactly
what we are, not gangs or uncontrollable
thugs raping, pillaging, and plundering. Our
clubs and our bikes are what we live for, and
anything else is just that, anything else. It goes
without saying that we've worked hard for our
reps and will do everything in our power to
maintain them. We do not apologize for a
damned thing. Not only that, if I knew that we
were going to live this long, I would have hit
life twice as hard. Some of the places I chose
not to go, I would most assuredly have gone,
twice. We are a group of complete individuals,
and I mean individuals. Every one of us has a

different reason for being who we are. The only thing we agree on is our love for the Club. That, and our love for motorcycles.

Another thing we agree on is the so-called biker gang experts. They have written books in collaboration with the biggest rat pukes you can imagine. They are the lyin'est bunch of jerks you've ever seen. They lie as much as most federal cops, and that's *really* saying something. One source for these experts is disgruntled ex-members who become rats. Most of these rats have had to make up most of the shit they say. Other than that, God only knows where they get their fantasies.

A *member of a 1%er club up in the far Northwest had to ap-*pear in court one time, and upon hearing his name called, he walked up to the bench. The judge was none too pleased with what he saw.

"Did I or did I not tell you to have an attorney present with you when you made your appearance here today?" asked the judge.

"Yes, you did, and I have one," replied the 1%er.

"Well, I don't see him," yelled the judge.

Not missing a beat, the 1%er took two steps to the right, changed his voice to a lower pitch, and replied, "I am the attorney representing this defendant."

The judge's face froze. After a few seconds, he regained his composure and said to the 1%er, "Have you ever heard the saying that anyone who represents himself has a fool for a client?"

"Oh, that's okay," replied the 1%er, quick-stepping to the left, in his own voice, "I also have a fool for a lawyer."

Then, after another long pause, the judge said sternly, "Well, what if I was to go and put your client in jail?"

Stepping back to the right, and in his finest lawyer voice, the 1%er answered back, "Well, then, Your Honor, I would be forced to file a writ of hocus-pocus."

While doing a stretch in Folsom Prison, one of our guys earned a living in the hobby shop making covered wagon lamps. They were really nice work, solid mahogany with all kinds of sharp detail, just excellent workmanship.

One day the Hobby Manager (who held the rank of captain) called our guy into his office.

"So, do you think you could have fifty-five of those covered wagon lamps ready by the end of next month?"

"Yup," said our guy, a man of few words but a devious mind.

This was during the California governorship of Jerry Brown. The state of California was sponsoring a governors' convention, and some bigwig state purchasing agent or some other such fool had contacted Folsom Prison and asked if they could help come up with some kind of gift for all the governors. That accounted for the request for our guy to produce fifty-five covered wagon lamps; one for the governor of each state plus Puerto Rico, Guam, Samoa, and God knows what other territories.

At this time, everyone in Folsom was allowed his own personal television set inside his cell. On the night of the convention, the local six o'clock news out of Sacramento (a short distance from Folsom) came on with the lead story. Governors' convention! They showed the facilities. Inside was a grand wooden oval table with fifty-five chairs, fifty-five water pitchers, fifty-five glasses, and fifty-five covered wagon lamps. The news anchor explained to the TV audience that the wagons were a special gift from Governor Jerry Brown to all the other governors, and that they were made and obtained from the hobby shop located on the grounds of beautiful Folsom Prison.

Tap, tap, tap, coming from the cell next door to the wagonmaster, from another one of our guys.

"You watching the news?"

"Yup."

"Well, it looks like you got 'em, brother, you sure as hell got 'em."

"Yup," came back the response.

You see, our master craftsman had long-dicked and ball-bagged every piece of every wagon, and that night every governor of every state and possession of the United States had 1%er dick and balls on his hands, and when they got home, so did their old ladies. They may not have known it (until now), but we sure did.

Viking Horde

It was 1967 in Lowell, Massachusetts. The little boy was just six years old, playing in the front yard of his grandparents' house. At first it was a low rumble from up the street. Then, with every passing second, the intensity of the rumble grew until it became deafening thunder. One by one, the boy's playmates disappeared, scooped up and dragged indoors by frightened parents until he was the last one standing, alone and outside.

The boy moved out to the edge of the sidewalk. First he saw the front two riders, then a mile-long procession behind them, motorcycles as far as the eye could see. And they kept coming. And coming. A Viking horde of black

Mike, "the little boy," many years—and many workouts in the gym— later. (Photograph courtesy of Michael Gervais)

and chrome choppers roared by, each rider more intimidating than the one before.

Instead of being frightened, the little boy was fascinated by the power and the fear that the riders held over the adults on his street, who at that very moment were cowering behind drawn shades. The boy stood there in all his innocence as the Club rode by, two by two. Some waved, some smiled, some looked on seriously with eyes straight ahead, all on their way to St. Patrick's Cemetery to bury a loyal fallen member.

Packin' Henry

Lake Don Pedro is a hard scramble, and not a very inviting place to have a run. East of Modesto, it's one of California's "best-kept secrets," tucked in the Sierra Nevada foothills in Gold Country. The MMA (Modified Motorcycle Association for you nonriders) used to hold an annual run at Lake Don Pedro at the same time as the world-famous Frog Jumps in nearby Calaveras County. The frog jumps were inspired by Mark Twain's story that put the county on the map back in 1865. For a long time, the promoters of the Frog Jumps weren't too friendly to bikers, so the MMA started its own run at the same time to create, shall we say, a more tolerant alternative atmosphere.

One year Cincinnati was there, and having

arrived early, he was sitting in the shade just inside the entrance, hanging out with friends. It was a little boring until Brother Ray from San Jose rode up. The two of them decided to take a putt over to the nearby small town of La Grange, a few miles down Highway 132. After a few drinks in the local tavern, Cincinnati and Ray took a walk around the tiny Western-style town for the express purpose of firing one up. Just up the corner, down the main street, they walked around behind a small hotel (circa mid-1800s). After a few minutes of puffing, things began feeling really weird, and it wasn't just the weed. The vibes were icy and damned spooky.

While both felt it, neither wanted to 'fess up, so they toughed it out a few more minutes. Finally, Cincinnati and Ray both looked at each other, said, "Fuck this," and headed back toward the main street corner. Ray was a bit ahead of Cincinnati, and when he turned around to say something, glancing past his buddy, he got this strange look on his face. Cincinnati turned around and saw this guy standing right behind him. He was wearing old-fashioned bib overalls and brogans. Cincinnati asked him, "Something you want?"

When the strange character didn't respond, Cincinnati barked back, "If you don't want nothing, then split."

Which he did, right in front of Cincinnati's eyes. Turning around, Ray was already heading around the corner real quick, with Cincinnati not too far behind. Stopping back in front of the bar, both were more than a little shaken up.

"Did you see that?" asked Cincinnati, hoping Ray hadn't.

"Yeah, and what the fuck was going on?"

The two headed back to the bar, acting real cool. Cincinnati asked the woman barkeep if La Grange was an old gold-mining town.

"No, sir, La Grange was a lumber town."

"So you ain't got ghosts here like they do in gold towns."

"Oh, yes, we do," she answered. "In fact, we have one who lives right next door, at the old hotel and livery stable."

Right then it was time for Cincinnati and Ray to get back to Lake Don Pedro.

The following May, it was back to Don Pedro. Bored again, Cincinnati decided to ride back over to the La Grange lumber town, hoping to find some human life. When he got to the bar, finding it closed, he started to make a U-turn. Cincinnati decided instead to sneak a peek around the corner at the old hotel. As he looked behind the tavern, he noticed they were tearing down the livery stable part of the old hotel where the bartender had said the ghost lived.

"I guess, Mr. Ghost, you're out of a place to be ghosting now," thought Cincinnati.

Shouting and shaking his fist, he yelled out, "Hey, Ghost, hey, Ghost, if you want to split, hop on!"

Cincinnati reached down and lowered his rear pegs when all of a sudden he felt someone—something—get on the back of his bike.

"Oh shit."

Cincinnati slipped into gear and started down the street. As he cornered the bike, it was wiggling and shaking so bad, he had to pull over and straighten out this damned ghost.

"Look, man, if you're going to ride with me, just sit still or lean in. Don't you wiggle or anything."

Just then a middle-aged couple on a full dresser rode by. They probably thought the day had been a little too kind to Cincinnati.

Back on the way to Lake Don Pedro, there were no more further problems, except now Cincinnati's thinking what the hell to do with his new friend. Spotting One-Armed Paul, he borrowed a tent, which he set up in the farthest corner of the campsite he could find, taking his new friend inside. It was there that he learned the ghost's name was Henry. He was nineteen (when he died?), and like Cincinnati, he was scared shitless. When Cincinnati and Henry talked, it wasn't through normal communication methods. It was more like telepathy, mind to mind.

Meanwhile, just about every road-dog friend at the run made their way up to "Henry's tent," some looking for a place for their ladies to crash. Cincinnati got a lot of weird looks when he said the tent was already full.

Now, Cincinnati has always been a stickler on rear pegs. When a girl gets off the back of his ride, the rule is she pushes them right back up. Cincinnati *never* rides alone with his pegs down. He says it looks like a taxicab driving down Madison Avenue with its doors open. So after the run, eight or ten of the mob were coming down the Mother Lode, Cincinnati with his rear pegs down, Izzy wanting to know why. Cincinnati pretended he didn't hear. As they glided into Oakland, everybody veered off onto their off ramps until it was just Cincinnati (and Henry) left.

"Now what?" Cincinnati was thinking.

Cincinnati decided to dump Henry off at the clubhouse. After situating him in the upstairs room (near the crap tables and pinball machines), Cincinnati jumped on his bike and headed for his pad in the Oakland Hills. He pulled into his garage just as his lady and kid were pulling up in the pickup. Cincinnati's old lady looked at his bike and immediately asked, "Who you been packing?"

"Well, baby," Cincinnati owns up, "I guess I brought a ghost back with me."

This went over big, so Cincinnati loaded his lady and daughter up in the pickup and drove down to the clubhouse. Once they got there, Cincinnati decided to stay downstairs.

"Go on," Cincinnati told his lady, "go on upstairs."

A few minutes later, his lady came back down with the hair standing up on her arms.

"Told you," Cincinnati said.

She gave him a look that could stop a train and told Cincinnati he sure as hell better get whatever-it-was the heck out of that clubhouse because he didn't seem any too happy. Cincinnati ran upstairs, and truly, Henry wasn't happy at all. That was when he found out that Henry flat-out didn't like women or kids. So Cincinnati drove his wife and daughter home and returned later to the clubhouse to pick up Henry.

Cincinnati lived in the Oakland Hills, and his garage sat right on the street, while the rest of his house was two floors down the hill, below street level. Very rustic, but homey enough for everyone concerned, including a ghost. Cincinnati's neighbor was hardly home, except for the occasional weekend. The neighbor's remodeled house had a hot tub, and Henry took to hanging there quite a bit.

Henry was not what you would call retarded, only a little slow. Cincinnati found out that he was orphaned at a real young age before dying at nineteen. The gentleman who owned the hotel/livery stable took him out to a farm he owned outside La Grange, where Henry was to be fed and watered like an animal by the man's wife. By the time he was ten, someone had partitioned off a front stall of the livery stable where Henry lived. The kids in town teased the hell out of him for being thick. The girls turned their noses up at him for being plain. Between the kids, the girls, and his benefactor's wife, that must have soured him on women and children forever. In 1896, Henry died before seeing his twenties.

Henry taught Cincinnati a lot about spirits, that there's no such thing as an evil ghost. Some apparitions are merely more aggressive than others. Not all spirits haunt. Some move around more than others. They travel the spirit highways to areas where ghosts prefer to gather, the Elysian Fields of spookdom, so to speak.

One night Cincinnati and Nic Tolbert were sitting in a bar when they decided to go on the whorehouse run that one of the chapters down south was holding in Beatty, Nevada. Cincinnati went home to grab a few necessities for the trip. He and Nic agreed to meet at his place. Cincinnati barely got there before Nic. Parking his bike in front of the garage, Cincinnati went down to the house just a few minutes before Nic pulled up. Now Nic and Cincinnati look a lot alike, especially at night, with long hair and full beards. Henry must have heard a bike, looked up, and saw Nic. Thinking he was Cincinnati, the ghost threw a paranormal fit. Nic noticed something weird right away.

"Cincinnati, you've got a spirit in here or something? All of a sudden these lights started moving around."

Half joking, Cincinnati told Nic, "Yeah, I've got a ghost living in my garage."

On October 20, 1991, Oakland suffered a huge firestorm. Five years of drought left a ton of dead trees and vegetation to feed the fire, with plenty of wind to fan the flames for miles. Hospitals in Oakland and Berkeley stayed on red alert. Hundreds of houses fell prey. The whole mess missed Cincinnati's place by three houses. After the evacuation, when people were allowed to venture back to their homes, Henry, freaked by the heat, went missing for several weeks. When he returned, he did so with a healthy curiosity about women. Cincinnati explained the best he could about the birds and the bees, the differences, the advantages, what to look out for, and the best way to handle the girls. Henry must have liked what he heard, because he confessed that in the heat of the firestorm, he'd met a lady.

"Cool," Cincinnati said to the love-struck spirit.

Several days later, Cincinnati was lying on the couch when his daughter came in from playing to announce that Henry and a lady were standing out by the garage. Cincinnati and his daughter went outside, and sure as shit, there they were. Henry introduced them both to Edith, his new girlfriend. At that moment, the most appropriate response seemed to be good-bye. Henry would be leaving Cincinnati's garage for good, and from that day on, he hasn't been seen again.

Now every time Cincinnati sees a passing bike with its rear pegs down, he thinks about Henry, hoping he's finally happy. While I can't say I believe in spirits, who but Cincinnati would pack a ghost like Henry?

Sister Teresa

Teresa is a law clerk and secretary by day. Her work space is decorated in a motorcycle motif. Framed pictures of motorcycles. Biker calendars. A wall of photos, awards, and proclamations. A Harley desk set. Her boss, one of the names on the door of the law firm, put it this way: He's hired more experienced people, but few work harder.

A relatively new Harley rider, Teresa has had her 1200 Custom Sportster since 1997. She loves the motorcycle lifestyle and reads *Easyriders* magazine. The sexy pictures don't bother her.

"A lot of us like to dress up in biker attire. We like the leather. We like men. We like to look good."

Sister Teresa, the beautiful, bike-riding, one-woman West Virginia fund-raising machine.
(Photograph courtesy of Mike Hall/PhotoGraphix)

Maybe it helps that Teresa is as good looking as many of the pretty models posing inside the *Easyriders* pages. But there's a lot more to Teresa than her good looks.

Teresa grew up around motorcycles in rural West Virginia, in a small town called Spencer. Her father bought his eight-year-old daughter her first minibike. Soon Teresa graduated to dirt bikes, and it wasn't until she moved to Charleston to become a paralegal that she felt the urge to begin street riding. Sensing a huge difference between dirt tracks and the street, she took the Motorcycle Riders Safety course, sponsored by the Motorcycle Safety Foundation, a nonprofit organization formed by the major bike manufacturers. After buying her Custom Sportster, she got involved in her local HOG (Harley Owners Group). That's when she became a joiner and a riding fanatic. It wasn't until she began riding motorcycles that her talent for organizing events and fund-raising came into play.

Soon Teresa felt the need to start an organization for women riders. She found that different women's riding clubs targeted different groups. Ladies of Harleys limited their scope to Harley owners. That was okay, but other organizations like Motormaids, Women in the Wind, and Women on Wheels seemed to include other riders, women who also rode Hondas, Suzukis, Ninjas, and so on. To put herself on the map, she needed ten members to get a

Women on Wheels chapter started. No sweat. Teresa enlisted thirty women who were ready to join by the time she submitted her first charter application. The calls continued, with some girls riding up to two hours to attend meetings. It wasn't long before Teresa's West Virginia WOW charter branched out into four separate regional chapters.

Women riding motorcycles isn't really an issue per se anymore. What is important are the specific concerns that many of the women talk about at Teresa's meetings. Women riders tend to be extremely safety-conscious, more so than men. Ninety percent of Teresa's members have completed their Motorcycle Riders Safety course. Another issue is that a lot of women feel isolated. Most find it hard to meet other women as companion riders. According to Teresa, some women feel intimidated walking into bike shops. What kind of riding gear should they choose? How do they learn to work on their own bikes? Most important, from an individual standpoint, what kind of bike is the best bike to start out with? Most women's stand on helmet laws tends to mirror men's, which is, safety gear should be left up to personal choice. Some of the more outlaw-oriented gals choose to ride only with men. Then there's the lesbian contingency, affectionately known in some parts as "Dykes on Bikes." Just like men, women on motorcycles represent a wide spectrum of riders.

With all the interest from women riders in West Virginia, what puzzled Teresa the most was the lack of on-the-street visibility she felt the local women riders had in her town. In other words, if all these women were calling her to join her WOW chapters, then where were all the chicks on bikes? She hardly saw them riding on the streets.

That's when Teresa organized the first-ever Women's Motorcycle Parade in her home state of West Virginia. The first was held in a huge parking lot used for the Riders Safety Course. Teresa hustled door prizes and sponsorships from motorcycle dealers. Soon she hit up other businesses, like corporate hotel chains, auto parts chain stores, and food companies. She booked guest speakers and live music. A short five-mile run on the interstate ended at the state capitol building in Charleston, where women riders were able to show the state and local politicians that they were an organized force. Teresa even got the governor to proclaim the day "West Virginia Women Motorcyclists Day."

Each year her parade draws hundreds of female riders. The run course was expanded to over sixty miles. When Teresa isn't working on her Women's Motorcycle Parade, she has branched out into a dozen other events for charities like Habitat for Humanity, March of Dimes, Make a Wish Foundation, the Red Cross, the Salvation Army, the Ronald McDonald Houses, various homeless shelters, veterans' organizations, and Sojourners, a local center for abused women and children.

"All of this has to do with my love of motorcycling," said Teresa. "That's what gives me my energy. I've been asked to do other charities, but if it doesn't pertain to motorcycling, then it doesn't spark my interest."

Teresa's charity drives extend to outlaw motorcycle clubs.

"I'll ride with the outlaw bikers on their poker runs. While I might not hang out in their clubhouses, I work with them so they'll in turn feel it's okay to give me support."

Like a lot of riders, what really drives Teresa is the power she feels on her bike and the admiration she gets on the streets, especially from other women.

"The one thing I hear a lot of women say is, 'I could never get on a bike like that and ride.' But you can. You just have to want to do it. I know women who ride Road Kings. Just like managing your life, it's not how large the bike is, it's how you handle and balance it. Women are capable of doing anything. All they need is a push, somebody to do the legwork. That's where I come in."

Fitzpatrick and the Iron Chopper

October 16, 1965, was a landmark date in the history of the Club. It was the day seven Club members and myself rumbled against eight thousand peacenik protesters and the Berkeley/ Oakland police at the Vietnam Day Committee (VDC) antiwar demonstration. As I wrote about the event in my first book, *Hell's Angel*, a few others and I stormed the crowd. When we first turned up, nobody was quite sure whose side we were on. But after we busted up a few college guys' heads, everybody figured out pretty quick that we were standing up for the men who fought and died in Vietnam, and not for the students in bell-bottoms and hippie beads. Being military vets ourselves, we were sick of the VDC Berkeley college elite looking down their noses at Oakland guys, like we were all a bunch of crackers from Alabama.

When we made the decision that afternoon to go down to the Oakland/Berkeley line and whoop some ass, we never really thought about any long-term repercussions. Looking back decades later, for all the shit we got from the antiwar left for speaking our minds with our fists and the end of our boots, a lot of soldiers who fought in Vietnam were inspired by our show of support.

Take the story of Army platoon sergeant Jim Fitzpatrick. Fitzpatrick loves motorcycles. His dad rode Indians when he was a kid, and the first real bike he owned was the 1949 Harley Panhead that "Fitz" bought for a hundred bucks. To him, there was only a thin blue line between bikers and soldiers.

Fitzpatrick's digital collage rendering of his Vietnam days with the 11th Armored Cavalry. (Photograph courtesy of James Fitzpatrick)

Fitzpatrick enlisted in the Army in 1958 as a patriotic eighteen-year-old. By the fall of 1965, at the rank of sergeant, Fitz was dispatched to Fort Meade, Maryland, where he prepared for active duty in Vietnam. Fort Meade was home to the elite First Squadron of the 11th Armored Cavalry Regiment, one of the first armored units to be sent over to fight in the jungles of Southeast Asia. Since it was early in the war effort, before they were sent over, platoon sergeants like Fitzpatrick had the luxury of handpicking their own posses from scratch. For Delta Company, Fitz chose the boldest crew of the hardest-ridin', hardest-fightin', and hardest-livin' cowboys he could find.

Fitz was as devoted to his new crew as they were to him. Being an NCO (noncommissioned officer), Fitz ate in the same grubby mess hall and slept in the same barracks as the other men. Fitz and his Delta Company trained hard with artillery, tanks, and APCs (armored personnel carriers). Soon they would venture into the jungles on search-and-destroy missions, where Fitz and Delta would be joined together on a 24/7 basis. Fitz's men nicknamed him "Pop," since at age twenty-five he was a few precious years older than the other guys in the tank company.

While in training, before being shipped off to 'Nam, Fitzpatrick and his squadron were placed on weekend alert to handle any possible antiwar demonstration flare-ups in nearby Washington, D.C. Curious, Fitz and a few of the guys slipped into D.C. dressed in civilian clothes to check out firsthand Washington's VDC demonstration, scheduled in tandem with several others across the country, on October 16, 1965. Knowing they would be soon sent off to battle, they were shocked to see demonstrators waving Viet Cong and North Vietnamese flags and cheering Ho Chi Minh on to victory over their own American forces.

Fitz and his now pissed-off cav troopers returned to base headquarters. They reported back to the other men sitting in the big recreation hall exactly what they had seen. Then somebody turned on the TV. The network news ran a report on that day's West Coast Berkeley/Oakland VDC demonstration. They talked about how a small but tough group of outlaw motorcycle riders caused chaos at the demonstration by stomping a bunch of protesters, even charging the main speaker, organizer Jerry Rubin. The dejected troops in the room cheered aloud when they saw film of the confrontation with bikers knocking in the heads of protesters.

Fitz's Delta Company mob never forgot what they saw. They looked at the antics of the Club as a ballsy show of support. Guys on their way to the front lines of Vietnam needed to know which Americans were behind them. A big impression was made on those young cavalry troopers that day—much more than the LBJ Great Society offered, citizens too busy working, making money, and buying houses and cars to show how much they really cared.

Fitz and the 11th Armored Cav boys were soon stationed in Xuan Loc, northwest of Saigon, between August 1966 and the summer of 1967. Most of the fighting took place inside the so-called Iron Triangle, not far from the Cambodian border, near a Vietnamese outpost called Ku Chi. Instead of riding Harley-Davidson Iron Horses in the Iron Triangle, Fitz's fighting squad rode fifty-two-ton Iron Horse tanks. In honor of the Club's stand that October day, Fitzpatrick named his M-48 A3 tank "the Chopper."

Delta Company held their mud with distinction, like the time when Fitz and his crew showed up at the battle of Suoi Cat: A Viet Cong regiment ambushed a small column of Americans. The call went out for reinforcements, and ten minutes later Fitz and the cavalry came roaring in with tank guns blazing. The Viet Cong opened fire as Fitz led the tank charge. The Chopper stood apart from the rest of the troop: a huge bulldozer blade was installed on the front of his M-48 A3 (complete with a twelve-cylinder continental vehicle

🦇 *Vietnam 1965. Fitzpatrick (the big guy in the middle) with "the Chopper" and some of the Delta posse.* (Photograph courtesy of James Fitzpatrick)

engine). The Viet Cong had felled a grove of large trees and spread them across the jungle road that led into the battle zone. The Chopper, brushing the trees aside like matchsticks, headed straight toward a Viet Cong rocket bunker. The weight of the tank crushed and collapsed the bunker site as Fitz backed up his rig, stuck the main gun tube down inside, and fired off a slew of deadly canister rounds. Many a Viet Cong foot soldier met his maker that day at the hands of Delta and the vengeful Chopper.

Battles like the one at Suoi Cat were typical of the escalating American artillery presence in 1966. At that time, the Viet Cong and

North Vietnamese units had never tangled with armored crews like the one led by the Chopper. In fact, most of the Pentagon planners were surprised that American tanks operated so well in the Southeast Asian jungle terrain.

It was the dawn of the Tet Offensive. The VC and North Vietnamese buildup from the DMZ was in high gear. With Fitz, the Chopper, and the 11th Cavalry in hot pursuit, they chased the Viet Cong infantry into Laos and Cambodia and watched helplessly as Charlie regiments slipped safely across the borders while both American armor and aircraft stopped cold. American command never did let the men finish the job. And their frustrations would only grow.

The Iron Triangle became home to a vast network of underground tunnels Charlie built beneath the jungles near the old French Michelin rubber plantation. Time after time, Delta Company would revisit villages like Ben Tat and Thou Cong, knock the shit out of Charlie, and resecure the area. Then they'd be ordered elsewhere. Three months later, the enemy would retrench those same sites and dig in with new bunkers, new ambush sites, and new weaponry. Command would send the armor and tank squadrons back in again and again until an ominous pattern developed: every skirmish grew tougher than the one before.

Fitzpatrick said good-bye to the Chopper and completed his one-year tour of duty. His military career lasted for a total of ten years. Fitz still looks back at Vietnam as the most exhilarating time of his life, but as with most Vietnam veterans, life was never the same afterward.

"Vietnam is always with me. A day never goes by when I don't think about it," admits Fitzpatrick. "No bungee jump, Mount Everest climb, or leap out of an airplane could equal the adrenaline rush of combat. I still remember the smoke, the noise, and the Chopper. Adrenaline drowned the fear. It was the ultimate hell-raising motorcycle run."

Today, Fitzpatrick teaches graphic arts and operates his own

 An original gas tank mural, Fitz's homage to all Vietnam veterans.
(Photograph courtesy of James Fitzpatrick)

design company. His minimural drawings honoring the plight of the Vietnam vets decorate Harley gas tanks, bike fenders, and cycle frames. Many vets ride customized choppers at bike shows that proudly bear his designs. Fitz calls his artwork cathartic and therapeutic; others call it patriotic. But Fitz still remembers that day in D.C.

"Sonny and his guys supported the 11th Armored Cavalry back in '65, and I don't care what they did before or since then. All I know is those American bikers stood up for a bunch of soldiers they never even knew."

The Last of the Unforgiven Few

The cops and the press usually don't believe me when I say the main purpose of the Club is to ride motorcycles together and have fun. But it's true; we are a club and we ride. End of story. For example, my chapter makes it a point to ride together at least one night a week, fly our colors, and hang out for a few beers afterward. Get the name of your club out on the street. Create some visibility.

A lot of bike riders wish like hell they had a good club to ride with. Earning your patch is not as easy as it sounds; you have to be voted in unanimously. You might be the greatest bike rider for miles around, but that doesn't necessarily make you the kind of member a club is looking for. You can't automatically move from

MC to MC. Private clubs regularly discriminate and only want the best. Like freedom, brotherhood isn't cheap or easy. Motorcycle clubs come and go, rise and fall—and when they're gone, they're gone.

Ask a disconnected bike rider like Woodminster.

Woody, as far as he knows, is the last member of a nomad club he calls the Unforgiven Few. The Unforgiven Few sprang out of the California prison system, and while there may be a couple of stragglers in some distant states, Woody has yet to regain contact with any of his brothers.

Woody was a born 1%er from Northern California, having been around motorcycles and bike riders all his life. As a kid he hung around bikers who belonged to some of the older West Coast clubs, like the Misfits, Satan's Slaves, and the Galloping Gooses. When he was five years old, Woody attended the funeral of Chocolate George, a famous member of our San Francisco chapter, and a loyal friend of the hippies in the Haight-Ashbury during the Summer of Love days.

When Woody was thirteen, the state of California declared him an incorrigible minor and moved him in and out of youth detention centers. As a ward of the state, he was ordered to report to a counseling center to hook up with a volunteer assigned to straighten him out. The center ended up pairing him with Skeeter, who rode with the Galloping Gooses. Skeeter was an active club guy and became a foster father/mentor to Woody.

Woody lived with Skeeter and his wife for a few years. What Skeeter didn't know about raising kids, he knew about making Woody a prospect. Being a teenager prospecting with a club like the Gooses was like being thrown into boot camp. Woody had to learn to tear down a motorcycle and put it back together in eight hours. On cold and rainy nights, Woody was the one who stood outside the clubhouse in the pouring rain, guarding the bikes during meetings. You could say Woody was rehabilitated; he knew when to shut up, keep his cool, and play out the scam.

One night Skeeter handed Woody an old Coast Guard ID. The picture on the ID looked nothing like Woody, but at least it said he was twenty-one. For the next year Woody and Skeeter drifted from bar to bar and hustled pool. Woody was the setup and distraction guy. He had a repertoire of rehearsed moves that could cause an opponent to miss a shot on demand. One night Woody was sitting at the bar by himself when the bartender asked him what he wanted to drink. Being an inexperienced (and underaged) drinker, Woody had no idea what to order and pointed to an expensive bottle of port over by the bartender's till. As the night wore on, young Woody got loaded on port and stuck all the drinks on Skeeter's tab. By the end of the night, all Skeeter had to show for a night of eight ball was a drunken teenager puking his guts out and a wrinkled ten-dollar bill. That was the beginning of the end of the Skeeter and Woody show.

A few months later, on another pool hustle run, Woody hooked up with a biker named Rebel. Rebel had just gotten out of the joint and rode with a nomad chapter of the Unforgiven Few out of California and Montana. Rebel, the club's VP, knew a streetwise bike rider with potential when he saw one. At nineteen, Baby Woody, as they now called him, became the youngest full-patched Unforgiven Few in the club's history. Woody never felt happier. The club made regular long-distance runs to the San Francisco Bay and areas north. He was proud of his new patch, riding with his brothers, and loved the idea of being a nomad. The open road was his and he felt welcome anywhere.

But a guy still had to earn a living. Woody bounced around and eventually got involved in a little interstate trucking and hijacking racket, which drew the heat. Well, actually, it was more than "a little" hijacking. Woody went on the lam with over 180 counts of receiving stolen property hanging over his head.

"I was pretty far underground," Woody admitted. "And I learned the only way to be a successful fugitive was to cut all ties with your past and consider your friends, family, and loved ones dead. I said good-bye to my MC. It was kind of frightening making that break. I remember the night I cut up my ID card and flushed it down the toilet. That was heavy."

Woody was a wanted man in hiding with a considerable reward on his head. His picture hung in the local post office. Rats, bounty hunters, and the FBI could lurk at any corner, so Woody hit the road and moved far away from the West Coast. He ended up in Ohio, not far from Columbus. As his new identity—Todd Sanders—he hooked up with a woman who operated a church of witchcraft and conducted seminars with cops to teach them the difference between witchcraft as a religion and criminal Satanism. Woody (as Todd) would help his girlfriend host the lectures and sit in the same room with all these cops—with a $50,000 price on his head.

Woody learned that the best place in the world to hide is often right out in the open. He got a straight job with a trucking company and for three years was doing a good job establishing his new persona. Then one day a couple of suits, a man and a woman, visited the office where he worked. It happened to be on a day when he was the only one there in charge.

The pair walked up to him and smiled. "Hey, Woodminster," they said. "How are you?"

"Who? Oh, I think you got the wrong office. I'm Todd Sanders."

"Todd, we're with the FBI. You got any ID?"

"Sure. Here's my driver's license, and I have my Social Security card, too."

Woody felt as if he had thrown them off the trail, but after letting them use a telephone to call their central office, the agents came back out. They needed to check to see if "Todd" had any identifying marks.

"I had my Unforgiven Few colors tattooed on my shoulder," Woody recalled. "When the FBI agent asked me to lift up my left shirt sleeve, I just said, fuck it, 'I'm the man you're looking for.'"

Woody was staring down the bore of a twenty-year sentence on several interstate hijacking beefs. In exchange for a lesser stretch (and not having to grouse anybody out) he pleaded out in Ohio, saving the state the time, money, and trouble of extraditing him back to the West Coast.

While inside the federal joint in Georgia, as an experienced trucking company manager, Woody ended up in charge of the food service warehouse. Being the survivor he was, Woody ran a few scams inside. Each night he brought his crew back to the warehouse and with their jackets slung over their shoulders, they left the compound with loaves of bread, cans of corned beef, and other goodies stuffed up inside their sleeves. Woody scored extra pocket money inside selling sandwiches to the inmates. Other than that, he managed to keep his nose clean and soon made parole doing a little over four years.

After his release from prison, Woody bought a new Harley and moved to the remote countryside where he lives today. Home is now a cabin on a highway that ends at the tip of a volcano, spitting distance from the Canadian border. Woody minds his own business and rides his 2001 Wide Glide up and down the winding mountain roads in the brisk northwestern countryside. He knows better than to stray across the border. The one time he tried to cross into Canada, the border Mounties jumped on his case, pulled his rap sheet, and ripped his bike apart.

Woody misses having a patch on his back, and 1%er blood still pumps in his veins. He admits there's a lingering feeling of emptiness, even on the Wide Glide. Woody misses the camaraderie of a club. So far, he has been unable to renew contact with the MC boys—Rebel, Baron, Preacher, and his other Unforgiven Few brothers.

Nowadays a disconnected Woody rides alone . . . and it's not by choice.

A Coupla Paychecks fer Ya

I **was sitting around my house on Golf** Links Road in Oakland one day with Cincinnati and the boys. We got a call, so I picked up the telephone.

"Is this Sonny?"

"Who needs to know?"

The voice sounded familiar. The Okie from Muskogee. Sure enough, it was him.

"Sonny, this is Merle Haggard. I just got a call from my office in Redding, and it seems Johnny Paycheck is on his way over there. They don't know what to do with him because I'm up here in Reno doing a show."

I knew what Merle was driving at. Johnny, one of my favorite outlaw country singers ("Take This Job and Shove It"), could be a handful sometimes.

Johnny Paycheck, longtime friend, outlaw country superstar, and champion of the workingman. *(Photograph courtesy of Michael Ochs Archives.com)*

We had a friend living just outside of Redding. I assured Merle that the situation would soon be in hand. I had Cincinnati call his pal on the horn.

"Hey, partner, do you know where Merle Haggard's office is?"

"Sure. Just down the road some."

Cincinnati explained the drill. "Paycheck's over at Merle's office and he's acting fuckin' strange. I need for you to go over there and find Johnny and hold him until we get to the bottom of things."

"I can be there shortly."

"Now whatever you do," Cincinnati warned his friend, "don't give him anything that comes out of a plastic bag."

We got back on the phone and called our Club brother Mickey, who said that he would send someone up to Redding to retrieve Paycheck and that we should get ahold of his road manager and find out what's cooking with this guy.

According to the tour manager, apparently Johnny was doing a series of concert dates, and a few days earlier, he had an attack of emphysema and was admitted to a Redding hospital. It was looking like he would be bedded down for some time, so they had canceled the rest of his tour. The band and crew had already flown home. The plan was that Johnny would fly back home a few days later, once he caught his breath staying with us for a while. The tour manager told me he would call the hospital and get back to us with the complete rundown.

On the callback, we heard the bad news.

Johnny had been placed in a room with an oxygen tube stuck up his snout and had been informed in no uncertain terms, no smoking. Johnny had smuggled some cigarettes and matches into the room inside his socks. Lying on his bed, he lit up a smoke and— *poof!*—singed all the hair off his face. After a heated argument with the staff, he was placed on the sidewalk in his hospital gown with a great big black cowboy hat on his head.

Cincinnati's friend hightailed it over to Merle's office, picked up Johnny, and paid back Merle's staff for the taxi bill. Then he took Johnny to his house. Brother Jesse from Sacramento was then dispatched to pick Johnny up a short time later and take him to Oakland, to Mickey's house for "observation."

The next day we got a call from Mickey. Could we do him a favor and come right over?

"But of course."

Over at Mickey's, past the dining room and into the kitchen, Paycheck was sitting in a chair with his knees up to his chest, holding an orange next to his ear, listening for something.

"What the fuck's his problem?"

"Don't know," said Mickey, scratching his head.

Mickey ran down all sorts of horror stories about the way Paycheck was acting. Shit, we knew Johnny was a hell-raiser, but never this bad. Cincinnati decided to take Johnny to his house. Maybe he was better equipped to deal with him.

Cincinnati scooped Paycheck up into his pickup and took him over to his house. Before he cut the engine, Paycheck bailed out of the truck and ran down the outside stairs, with Cincinnati in hot pursuit. As Cincinnati rounded the corner to his front door, Paycheck was pinned to the wall by Stryder, Cincinnati's Doberman. Pulling the dog off Johnny, Cincinnati finally got him into the house. Baz, a brother from Omaha, was visiting, so the two of them set up a round-the-clock Paycheck watch.

But Johnny continued acting crazy as a bedbug, and neither Cincinnati nor Baz had the foggiest idea what the fuck was wrong with him. Things seemed serious. Johnny was climbing the drapes, smacking lamps, and talking nonstop craziness. His actions were making no sense whatsoever. Cincinnati called me up, and usually I can handle anything. This time I had no clue what to tell Cincinnati.

Cincinnati split in search of some kind of downer in hopes of at least getting Johnny to sleep. On the way back, he stopped at a convenience store, and as he passed the magazine rack, he spotted an issue of the glossy mag *High Times* with a screaming headline, "Cocaine Poisoning." Grabbing the magazine, Cincinnati drove home and read the article, and it fit Johnny to a tee. He followed the article's instructions and soon gained control of the Mad Paycheck situation. Johnny slept for several days, and while he was dozing, we made arrangements to fly his outlaw ass back to Atlanta.

The next day Cincinnati (with his lady) and I took Johnny to the San Francisco airport. Inside the terminal, Johnny wanted Cincinnati to stand in line for him, but he soon found out quick that we weren't taking to any of his star treatment bullshit. After we got his ticket, Cincinnati's old lady talked to the stewardess and the airline folks agreed to walk Johnny to his plane, put him on board, then see to it that he made his connection in Dallas/Fort Worth, then back to Hot 'Lanta.

We walked Johnny and Cincinnati's lady past security and up the departure ramp, and then headed for the bar. After a couple of drinks at the bar, we looked up, and there was Cincinnati's lady back with Johnny. Somehow he had fucked around and missed his plane. Cincinnati told his lady, "You had better do something and do it quick, because I am not leaving until this little sucker's cowboy hat is forty thousand feet in the air."

She came back quick with a nonstop ticket to Atlanta. Now all we had to do was call Johnny's wife and tell her to meet up with him in Atlanta. Sure as shit, he would be on that plane.

As we sat at the table in the bar waiting for the flight to take off, Johnny's head was resting on his hand, with his elbow on the tabletop. He must have been wasted and dozed off, because his elbow slipped off the table and his head hit the tabletop with a loud, hard crash. Everybody in the bar looked over as Johnny jumped up hollering, "What did I do? What did I say?"

Johnny must have fallen asleep, dreaming that he'd fucked up a member of the Club and that one of us had smacked him. We assured him that had he been smacked by us, there would be no doubt it was real and not a dream. A few minutes later, Johnny hit the friendly skies, and frankly, while we love him and his music like a brother, we were quite happy to see the back of his ten-gallon hat.

The next time Johnny Paycheck was touring on the West Coast, he visited us again. We were sitting around my Golf Links house bullshitting with Paycheck when there was a knock at the door. Someone opened the door and a friend of ours (and a huge Paycheck fan) was let into the house.

"Hey, Johnny, come outside." The guy had a grin from ear to ear. "I got something just for you."

We all walked outside, and right in front of the house sat this nice little Sportster. It's got a 16 laced up on the back, a nice 21 on the front, a killer paint job, and lots of chrome. The guy could hardly contain his excitement. Our buddy fired it up and motioned over to Johnny.

"It's for you, Johnny."

Up until that afternoon, Johnny was talking shit about motorcycles. He'd say stuff like, "Oh, I got to get me another bike."

We all knew that the only time Johnny had ever been on a bike was on the back of Mountain's, when he would pack Paycheck around town once in a while.

Well, the Sportster was running, and Johnny threw a leg over the Harley. *Hurrrrum, hurrrrum,* it went. Johnny was running it up to maximum rpms and holding it awhile. He looked down to his left and spotted the shift lever when a look of "I wonder what that could be?" came over his face.

"Hmmm."

Maximum rpms, toe, touch, *wham,* the fucking bike leaped straight into the air like a wild stallion, coming down on its rear wheel, which immediately kicked up a cloud of smoke, slamming the rear end to the left and then to the right. Johnny's hanging on with one hand, his legs sticking straight out behind, locked in, holding the bike at wide-open throttle.

Paycheck's bucking Sportster shoots straight across the street, heading for Crazy Earl's garage. Earl was in there wrenching on his bike when he heard this entire shit going down. We saw Earl's head poke out between the vertical barn doors of his garage. Spotting the screaming Sportster heading straight for him, he pushed the door out in front of Johnny to break his attack.

Boom! Paycheck hits the door, ricochets off it, and rolls downhill into the field next door to Earl's, plowing up a big furrow in the ground with dust and weeds and bushes and all kinds of shit blowing up in the air.

Earl calmly shut the door and disappeared back to the safety of his garage.

Luckily, the bike died.

Back across the street, we were all on the ground, holding our guts, laughing so hard some of us were puking and pissing our drawers. I yelled out, "You stupid motherfucker, what the fuck is wrong with you . . . I thought you could . . ."

While we were trying to catch our breath, here comes Paycheck, with dirt all over him, weeds, sticker bushes, and rocks stuck to his hair and clothes. His nose is bleeding, road rash and blood all over him, his big old black cowboy hat just a memory. Holding his head in one hand, he looks at me, takes a ragged breath, and asks, "Sonny, y'all think I can keep it?"

Lone Wolf and the Bad Boys Crew

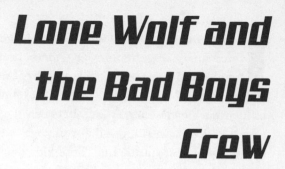

Some of the most progressive chapters of the Club are the European charters, based in places like Germany, Scandinavia, Great Britain, Italy, Holland, Belgium, Switzerland, and many other countries. The first Austrian outpost of the Club was formed in 1975 in Vorarlberg. Vorarlberg is on the southwestern tip of the country right near the Swiss and German borders, while Vienna is on the other side, in the northeastern region near Slovakia and Hungary. With Vorarlberg in place, the fledgling Vienna chapter collected a group of prospects from an existing Austrian club formed in the early 1970s called Satan's Serpents.

n 1970, before our two international chapters in Austria were voted in, young motorcycle riders in Vienna were already caught up in the outlaw biker frenzy. Motorcycle movies like *Hell's Angels '69*, *The Wild One*, and *Hell's Angels on Wheels* heavily influenced Viennese street kids between the ages of eighteen and twenty-five to start riding—the same ages we were when the Oakland chapter first started up. These kids read about how tough we were and saw pictures of wild bike riders on their Harleys. They longed to be just like them—fighting in the bars and going on cross-country runs. Soon they scraped together enough schillings to buy motorcycles and form their own clubs with English names like Bad Boys Crew (*Böser Buben* in German), the Savage Skulls (*Fetzenschädeln*), and the Vienna Beasts (*Die Wiener Viecher*). But instead of guys called Goose, Hi Ho Hal, or Big Del, the Bad Boys Crew had members called Redhead, President Peter (Vienna's first 1%er), Uwe, Sioux, Guzzi-Chris, Ventil-Kurt, Franz, Speedy, Moped Franz, Karl, and Schwammer.

Lone Wolf, a bored nineteen-year-old university dropout, rode with the Bad Boys Crew for six years, starting in 1972. At any given time, the Bad Boys Crew swelled from twenty-five to thirty members. Compared to today, Wolf remembers those MC days as much more innocent. But the Bad Boys Crew raised enough hell to keep the cops off balance. In an attempt to imitate our Bass Lake runs, about seventy bikers (including thirty Bad Boys Crew) rode in on a pack to converge on a small village in Styria called Perbersdorf. And just like in America, the country sheriffs called out for help from neighboring towns. Things got tense when the law showed up at their bonfire campsite. The idea was to pressure the young riders to split. The Bad Boys Crew and their friends refused to leave town, and outside of a couple of cops getting punched, it turned into a standoff. The cops didn't really know what to expect. The Bad Boys Crew had come to party, and weren't willing to let the cops completely off the hook. A couple of young members stole the blue siren

Lone Wolf in 1972 on his chopped Honda 750 roaming the streets of Vienna as a member of the Böser Buben (Bad Boys Crew).
(Photograph courtesy of Lone Wolf)

lights off a couple of their squad cars. Outside of the loud roar of their bikes and a bunch of drunken Viennese riders, things never reached the boiling point on the level of a Porterville or Hollister. As the Crew rode out of the village, the boys deposited the stolen sirens on the doorstep of the police station.

Back in the 1970s, Western Europe still had a lot of catching up to do in terms of bike culture. Early motorcycle MCs like the Bad Boys Crew wore their colors as arm patches, as opposed to three-piece patches on their backs. Harley-Davidsons were much harder to come by in Austria. Aside from those left over from U.S. service-men in World War II, new Harleys trickled into Vienna at the rate of a couple dozen per year. As a result, the younger clubs rode chopped versions of Honda 750 ccs (Lone Wolf's ride), Norton Commando 750s, Suzukis, Kawasakis, and some Triumphs. It didn't really bother guys like Lone Wolf that Harleys were so scarce; besides, it was during the AMF period, and those bikes not only leaked oil like crazy, they lacked the gas efficiency of Jap and Brit bikes. Mileage was an important factor when you took into account the European price of fuel. Later on, a few riders did pick up used Harleys at American police auctions in the States and shipped them back to Austria to have them rebuilt.

In typically precise, Teutonic style, the city of Vienna is sepa-rated into numbered districts. The Bad Boys Crew had their club-house in the 18th District, while the club hung out in a little café bar called Café West in the 7th District. They later frequented another rough, dark cellar joint called the Wallensteinkeller along the river Danube in the 20th District. The Wallensteinkeller played American blues on its jukebox, served Yankee hamburgers, and was an easy hangout for the black market and criminal element of Vienna. There was an amusement center in town called the Prater. While today it's a popular tourist center in modern Vienna, that's where some of the Bad Boys Crew originally rode the streets and hung out, protecting local pimps and prostitutes. Vienna was a fun biker city where riders could drink strong Austrian and German beer, smoke a little reefer, and party.

The heaviest clashes between the early bike clubs in Vienna went on between the Bad Boys Crew and the Savage Skulls. While the oldest members of the Crew were twenty-five, the Skulls were a slightly younger posse (ranging between sixteen to twenty) and more cocky. Pretty soon the Skulls caused enough aggravation to become a sore spot for the Bad Boys Crew. It all came to a head when the young Skulls knocked off a gas station. This drew heat on

Lone Wolf and his Crew brothers, since the gas station fell within the Bad Boys' clubhouse turf in the 18th District.

It was soon time to come down hard, so in retaliation one winter night, the entire Bad Boys Crew drove up in their cars outside a dive called the Paradiso, the Skulls' hangout in the center of Vienna. Stashing their wheels, the Bad Boys pulled out the necessary baseball bats, knives, chains, and knuckle-dusters in order to finish the Skulls off once and for all.

That night a winter storm had left a frosted sheet of ice all over the road outside the bar. As the Crew readied to rush the bar, yet *another* rival MC approached the bar from the other end. It was thirteen members of the Vienna Beasts, who also had a beef with the Skulls and were ready to do battle. Instead of ganging up on the Skulls, the Beasts and the Bad Boys Crew exchanged their own threats and charged each other. The outbreak resembled a demented but comical hockey game more than it did a rumble. Bikers swung bats and chains at one another, falling on their asses more often than making contact.

Lone Wolf's cheek caught the tail end of a flying bicycle chain. And although the Crew outnumbered the Beasts two to one, it was still hard to land a punch and stay standing. As the Crew slipped, crawled, and slid around, fighting as much on their knees as on their feet, there were as many cracked kneecaps and broken legs that night as there were broken noses and busted jaws. The "ice rink" out in front of the Skulls' bar was soaked with blood as the police rounded up both clubs. All the while, the Skulls continued drinking without realizing what had gone down outside.

Wisely, the Bad Boys Crew waited until spring for things to cool down before they settled the score again with the Skulls. This time the Crew placed spies inside the Paradiso to pick their moment. Instead of brass knuckles and baseball bats, the Bad Boys had sawed-off pool cues and snap knives up their sleeves as they entered the tiny bar en masse. That night twenty-five Bad Boys were on hand to fuck up a dozen Skulls. The bar was so small, they had to duke it out and mash heads in shifts, half a dozen inside at a time. The Skulls got busted up pretty badly as the Bad Boys made their getaway after a waitress called the police.

During the mid-1970s, the biker scene in Vienna really began to toughen up. Satan's Serpents came onto the scene and quickly graduated from Kawasakis and Triumphs to full-blown Harleys. Their numbers grew, as did their reputations. The Serpents began wearing three-piece patches with bottom rockers, taking a more aggressive 1%er stance against the other clubs. Soon the Bad Boys Crew had pretty much seen their day, sitting nervously in their Café West

The Lone Wolf today with Viennese motorcycle babe.
(Photograph courtesy of Lone Wolf)

hangout, watching the Serpents ride by in meaner and larger numbers. It became obvious that Vienna was now Satan's Serpents territory.

The Savage Skulls disbanded shortly after the Bad Boys Crew took them out. Some of the guys from the Vienna Beasts moved on to Bavaria and formed another club called the Road Eagles. By 1978, the Bad Boys Crew, along with Lone Wolf, disbanded their club and deferred control to the Serpents. It was right around that time when the Vorarlberg chapter of the Club came in, took the Serpents under their wings, and eventually expanded into Vienna.

When it isn't snowing too hard in the Austrian country-side, Wolf rides a new Harley, recently chopped and customized by a Vienna Club guy. He has a family now and a cabin outside of Vienna, his MC hell-raising days long gone after the demise of the Bad Boys Crew. Lone Wolf, an avid 81 Club supporter, enjoys hanging out at its nightspot in Vienna, appropriately called The Big Red Machine. As Wolf kicks back a few beers, he recalls a bygone era when a Lone Wolf showed his teeth and young clubs rode the Vienna streets and paved the way for a new generation of 1%ers . . . and a rougher, tougher new ball game.

Joey Keeps His Promise

Life was pretty damn good for
Joey. He'd just bought his first Harley,
a 1968 Sportster. He had a few bucks
in his pocket. He was juggling a few
girlfriends. How could things get any better?
But just when Joey had life by the balls, his
ride broke down in the New York suburb of
Westchester. There he was, stuck on Route 9,
well over an hour's ride from his pad back in
Manhattan.

Joey was a new rider, not a wrencher. He
suddenly felt that cold shot of reality: ma-
rooned on the side of a busy highway without
a clue as to what to do. Shit like this always
happened to somebody else, not Joey. He didn't
know a single soul in all of Westchester County.
His buddies in Little Italy didn't have cars.

(This was before the days of cell phones and express emergency road service, so Joey was a sitting duck, shit out of luck.)

Just as he was ready to give up the ghost and hoof it down to the town center, an older hippie-looking fellow rolled by in his beat-up pickup truck. He parked his truck close behind Joey's dead ride. He wore dirty overalls and spoke with a thick Chowderhead accent.

"Looks like you could use a hand."

Joey smiled and shook the guy's hand. Yep, he was broken down. As the hippie reached for his toolbox from out of the truck, Joey panicked a little. "Shit," he thought to himself, "this guy looks like he could barely fix himself a ham sandwich, much less monkey with my bike."

But Joey was dead wrong. The hippie figured out Joey's problem pretty quickly. Something was funky with the spark plug cap. The spark plug cap was a little loose (probably from vibration). He jiggled the plug cap a little, then pulled it off, crimped the connector a bit, and snapped it back on the plug. That's all it took.

Varoooom! The bike started right up again.

"There you go," he said as he put his tools back inside the box.

Relieved that he was going to make it back home okay, Joey pulled out his wallet as the hippie walked back toward his truck.

"Damn, partner," said Joey to the Good Samaritan. "I owe you something for fixing my bike. How much?"

The hippie waved his hand. "Tell you what. Make me a promise. If you ever see a motorcyclist stranded, stop and help him out."

Joey promised he would, thanked the Good Samaritan profusely, and they both rode their separate ways.

Sixteen years passed. *Joey was driving his pickup truck* down the turnpike from just outside Manchester, New Hampshire, where he was working a construction gig. His buddy and workmate Rick was in the front seat with him. It was a Friday afternoon, the end of a long, tough week on the site. Joey

and Rick were looking forward to driving down to Boston, partying with some pals and gals, and letting off some steam for the
weekend.

As they took the Route 89 exit toward Boston, Joey saw a motorcycle stopped on the side of the freeway. A burly guy with long
hair and a full beard stood over his troubled chopper. Although it
had been years, Joey remembered the promise he had made to the
hippie and pulled his pickup over.

The biker's name was Dawg. He told Joey that he was on his way
back home after a big bash in Canada with a bunch of his bike-
riding friends and got separated from the pack. Apparently, on the
way home, Dawg had tried to put the moves on a waitress in a coffee shop outside of Concord. She was fine, and he figured she was
his for the asking. Meanwhile, Dawg's riding buddies had seen this
movie over and over. They weren't going to stick around for the
same old ending and watch Dawg strike out again with this chick.
They were moving on. So the pack started up their bikes and roared
off without him.

Unfortunately, the guys were right again. The waitress behind
the counter was another dud. In fact, she was downright rude. Now
here he was, broken down on Route 89, solo.

Joey confessed to Dawg that he was no mechanic, but he and
Rick *could* help load Dawg's bike into the back of the pickup and
drop him off at a shop or somewhere close. Dawg accepted Joey's
offer and soon they were back on the road, heading toward Boston.

"So, what's up for you guys this weekend?" asked Dawg, initiat-
ing small talk and friendly chat.

Joey told Dawg about the rough week they'd just had and that
they were looking forward to some hard partying with friends in
Boss-town. With that, Dawg smiled broadly, pulled a fatty out of his
breast pocket, and fired up.

"Well, let the party begin! I've got this amazing Vancouver weed
from one of my guys in Canada."

Dawg's reefer made the rounds inside the cab of the pickup as all three guys got pretty toasted. As they approached the Boston area, Dawg asked if it would be okay if Joey and Rick dropped him off in Lowell. It was an extra half hour's drive out to Lowell from where Joey was planning to go. But at that point, Joey and Rick were so zonked on Dawg's weed, hell, they would have driven him to Miami had he asked.

"No problem," laughed Joey, who had the "slit-eyed grins" from the stranger's pot. "Just tell us where to go."

As the three men rolled happily down the road, Dawg turned out to be a real down-to-earth cat. Soon everybody was throwing out their best bullshit. They talked about bar brawls, hell-raising motorcycle tales, babes, and the Red Sox. Time flew by fast, everybody was still buzzing, and the next thing Joey knew, they had arrived at Dawg's doorstep.

Joey and Rick gulped when they found themselves in front of the Lowell clubhouse. As they unloaded the bike from the back of the truck, Dawg could see that Joey and Rick were a bit freaked out, maybe even a little scared. Manchester, where they had just worked, was a big biker town, and the Club had a notorious rep among the work crew guys there. But never mind. Dawg invited them to stick around.

"C'mon in and have a beer and some barbecue," he said, slipping on his colors inside the clubhouse. His vest pocket had a president's patch sewn on it. "You guys are with me, so if you're really into partying hard, I suggest you hang out here for a while."

The other members were glad to see Dawg back in one piece, out of trouble, or at least not sitting in some small-town jail somewhere. And the members treated Joey and Rick with respect. After all, they had rescued their stranded brother. Joey and Rick toured the historic clubhouse and checked out the banners, plaques, and mug shots of the members on the walls. Within hours the place was a circus, jammed with beautiful leathered chicks, lots of brew, smoky

BBQ, tons of members and their friends, a live blues band, and a pair of foxy strippers. Joey phoned his mates in Boston, screaming over the noise of the party. There'd been a sudden change of plans. Joey and Rick were stickin' around Lowell for a couple days. They were much happier hanging out with these guys, touring the town, partying, and riding some amazing customized bikes, courtesy of their new friend Dawg. After two days, it was time to head back to Manchester. It had been an amazing weekend and a fitting reward for a bike rider who kept his promise to obey the Code of the Road and *always* help a stranded bike rider in need.

Kaye Keeps on Truckin'

In a small town just a half hour west of Salt Lake City, Kaye and her husband, George, owned and operated a service station, garage, and towing business situated on a busy highway. The two had certainly seen their share of the aftermath of car and bike wrecks. Hardly a day passed when a truck wasn't dispatched to clear up an ugly accident on Interstate 80. George and Kaye worked hard, and business was good, so eventually the couple branched out, opening a second operation a little over a hundred miles west in Wendover, an even smaller town next to the Utah/Nevada border. Most of his life, George had been a Harley rider, while Kaye was *usually* content to ride on the back. Then, in 1991, Kaye's mother passed away, leaving a few dollars behind. The couple talked about it and both agreed it would be a good idea if Kaye spent her small inheritance on something meaningful that might memorialize her mother.

The next day when Kaye left on her shopping trip, George assumed she would make it over to the jewelry store and pick out a ring or a diamond necklace. But instead of rings and baubles, Kaye had something else in mind. Driving past the jewelry store, Kaye turned the truck toward Salt Lake City.

Walking into the Harley dealership, there it was, a new 1991 Heritage Softail, parked right in the front of the store. Except for the "SOLD" tag that hung from the handlebars, this was exactly what Kaye had in her sights. Finding the right Harley is usually an exercise in good timing, sometimes taking weeks, months, or, in some towns, up to a year to order the right ride. However, on that day, Kaye got lucky. The guy who had ordered the bike hated the color green. Some bikers thought it was a bad-luck color. To Kaye, factoring in the chrome and the bike's sexy rake and stretch, its color didn't matter. According to the shop manager, if she wanted it, hell, the Heritage was hers. Kaye whipped out her checkbook. Soon her late mother would be memorialized in a way few in the family ever expected.

When Kaye returned to the station with the Heritage, George's jaw nearly hit the concrete garage floor. After years of riding on the back of *his* bike, Kaye was joining the ranks of women ready for their own bikes. With a little practice (and after taking a safety course), the Heritage and the highway came in handy for Kaye when it came to battling the stress involved in running *two* service stations, *two* towing companies, and payroll for *two* convenience stores, not to mention the needs of her family.

It was a warm Wednesday morning in January when Kaye and George had a hell of a fight over not a hell of a lot. But it was enough for Kaye to want to call it a day, grab her cold-weather clothes and gloves, and hop on the Heritage. A little wind in her hair would blow off the steam, so down the road she went. How far she'd ride, Kaye didn't know, nor did she care. In the huff she was in, she might just putt it all the way to Wendover, check in on the guys there, and ride back by nightfall. By then, she and George would be back in sync, maybe catch a little Italian food and vino for a late dinner. In the pocket of her leather jacket, Kaye felt the bulge of her cell phone. Getting away from the telephone was part of the reason she rode, but something told her not to chuck the plastic beast into the brush along the highway.

Ten miles down 80, the January chill started to make its presence felt. The Heritage was running smooth, so if she slipped into her winter gear, the next hundred miles might be nearly perfect. Signaling, Kaye pulled to the side of the highway. Pulling her clothing from the saddlebag, she slipped on the light Gore-Tex travel gear while out-of-town skiers and vacationers whizzed by in their SUVs.

Kaye had just put her gloves on and was standing next to her bike. Just then an eighteen-wheeler came barreling past. Kaye could feel the breeze the huge semi had generated, except that wasn't all the truck left behind. Two giant double wheels spun off its axles and were heading completely out of control toward the roadside . . . toward her! The first wheel shot about ten feet in the air, whizzing just over Kaye's head. The second, however, broadsided Kaye's Heritage, hitting her fuel tank. The sudden impact of the wayward wheel knocked Kaye into the air and into the ditch, and hurled her bike twenty-five feet. Lying flat on her back, Kaye watched her beloved Heritage bounce off a telephone pole before landing a few feet from her in the ditch.

Kaye remembered the cell phone in her pocket and dialed 911, but all she could worry about was her bike.

"I figured my legs would heal, but if they couldn't fix my bike, no way could I wait a year or so for a new Harley."

Back at the station, George watched and heard the patrol cars screaming toward the scene of an accident down the road. Instinctively he jumped into the wrecker and headed toward 80. When he got to the scene, he was shocked to find Kaye lying in the ditch arguing with the paramedics about whether or not they were going to cut off her new leather jacket. The medics obliged and smartly backed off. By the looks of it, Kaye was now even more pissed off than when she left the house.

All George could do was watch the commotion, and when the driver of the truck came over and offered to apologize to Kaye, George warned him to keep his distance. He'd never seen Kaye this fired up. The last thing he or the police needed was an assault charge.

The ambulance trip to the hospital revealed only cuts, bruises, and a knee injury. They pronounced Kaye damned lucky to be alive. If that wheel had hit her another six or seven inches higher, they reasoned, Kaye might be playing checkers with Elvis.

In the final analysis, Kaye underwent knee surgery to repair torn cartilage. As for the bike, it lived, sustaining about $11,000 worth of damage and an entire winter holed up in the repair shop. That hurt more than the surgery, and it was hard to tell which bent frame seemed more important to Kaye, hers or the Heritage.

George would try in vain to get his wife to sell the Heritage, but Kaye and the bike had been through too much together with the accident and all. Despite the scrapes and bruises, she still felt strangely protected, as if the memory of her mother was still watching over her.

A couple years and 26,000 miles later, on the anniversary of her mother's departure and gift, Kaye entered her Heritage in a local bike show at the Salt Palace. Next to the Softail sat the original dented gas tank. It serves as a reminder that luck and faith play a big part in riding and surviving on your motorcycle. And as we all know, when it's your time to go, you're gonna go. Yet until that time comes, Kaye figures she might as well have a good time riding. Some bike riders say: *Ride to live, live to ride.* Or as I say: *Ridin' high, livin' free.*

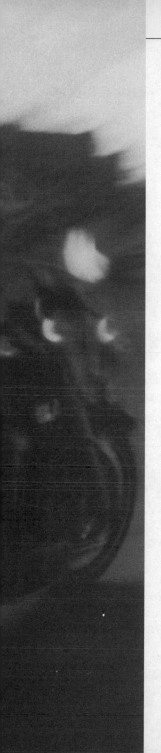

Ranger Holds His Mud

Ranger rides with a small club he formed five years ago. The Squires MC has fewer than a dozen members—all close blood brothers—within its small chapter. Realistically, the Squires (*not their real name*) fall into the category of "10%ers" as opposed to 1%ers. But flying your colors on the open road these days as a tiny single charter can be a tricky proposition. Turf politics can be a major reality when riding and hanging out with your MC brothers. As a small club, you are a fish in shark-infested waters. Sometimes it's eat or be eaten.

When Ranger formed the Squires, he approached the two major MCs with chapters in his 'hood to let them know he and his boys were around. In the interest of not wanting to

Ranger kick-starting his pride and joy.
(Photograph courtesy of Ranger)

Ranger's ride, built from scratch, parked outside the Moonshiner. Kennedy's custom frame, EVO motor, four-speed trans, Lepara seat.
(Photograph courtesy of Ranger)

shake things up, rather than wearing Nomad bottom rockers, the Squires elected to use the word "Renegades" on their lower patches. Ranger got the all clear sign from both large 1%er clubhouses, and the Squires rode the fine line. But Ranger, whose brother and father also rode with traditional MCs, knew that trouble could be just around the corner. Getting the big boys to give you the thumbs-up was one thing; getting their support clubs to respect your independence was another barrier to cross.

One Sunday, Ranger, his old lady, Melanie, and his club brother (and Squires VP) Festus sat around Ranger's crib bored. There was no hockey game on the box, and the rain had cleared up just right for the weekend. Ranger decided it was prime time for a scoot out to the country. Festus suggested they hit a well-known biker watering hole called the Moonshiner. It was an hour's mountainous excursion away, where green hills and sprawling farmhouses replaced the grimy city street corners and graffiti-tagged buildings of home.

Ranger (packing Mel) and Festus threw on their colors and revved out to the Moonshiner. As they approached the front entrance of the rural tavern, they recognized most of the Harleys parked out front. Most were independent riders, some HOGs, as well as other small clubs the same stature as the Squires. Off to one side, a long row of bikes obviously belonged to the Mainliners (not their real name either), a support group of one of the big clubs. There must have been a couple dozen Mainliner bikes, but Ranger and Festus didn't pay them any mind. The Moonshiner was a peaceful, neutral hang. Over the years neither Ranger nor Festus had ever thrown one punch in this place. As the three entered the bar, there they were, the Mainliners, off to one side, keeping to themselves.

Ranger ordered the first round of beers and started up a conversation with an indie rider he knew sitting a couple of bar stools down. Then Festus tapped Ranger on the shoulder. The president of the Mainliners wanted to speak to him, and Ranger immediately had a bad feeling. He motioned over to Mel and suggested she ask the bartender to call her a cab. There might be some trouble at the

Moonshiner. But being the straight-up old lady she was, Mel stood by her man. She was staying put.

Ranger, as the founding president of the Squires, understood the situation he was about to face. It wasn't the first time. But that didn't make it any easier. Of the two big 1%er clubs, the Squires' bikes displayed support stickers for Club A, while the Mainliners clearly aligned themselves with Club B. Ranger knew one thing: it was time to hold his ground and handle the situation with the right combination of diplomacy, balls, and guts. At this juncture, he honestly didn't know whether he was walking into a potential rat pack or if he could talk this thing through.

Before Ranger, Festus, and Mel approached the club president and his mob, they counted about twenty associated members scattered around the bar. Ranger got right to the point and stood toe to toe with the Mainliners' president.

"Say your thing, man." At that point the whole club formed a tight circle and closed ranks around Festus and Ranger.

"Is it true that the Squires hang out at that strip club, the Twister-A-Go-Go?" asked the prez.

Ranger nodded. "What about it?"

"One of my guys was in the Twister a few days ago and he was disrespected by one of your boys. This is a bad fucking thing, and I ain't gonna allow it."

"Then tell me this." Ranger pointed around the room. Each Mainliner had a belligerent look on his face. "Which one of your guys was disrespected? Tell him to step forward. I wanna talk with him."

Prez paused, then laughed.

"Well, he ain't here right now. But that doesn't solve our problem, man."

Right then Ranger knew the guy was bullshitting. Ranger had a choice to make: either back off and punk out of this messy scene or stand up to these assholes, even if it meant getting carried out sideways. He pulled out a Squires MC club card and handed it to Prez.

"Tell you what," said Ranger. "Here's my phone number. You get your guy, and I'll get all my brothers together, and we'll do a lineup. But I will say this: If your guy can't identify my guy, then your guy is going to the hospital. Understand? Something tells me if we weren't looking at a twenty-to-two standoff here, we'd never be having this fucking conversation."

The room grew tense and silent. Ranger and Prez stood eyeball-to-eyeball.

Prez blinked first. "Listen, there's no need to do a lineup."

Ranger knew then that he had the upper hand in this confrontation.

"Then why call me over here? To see if I'd run scared? Look, do what the fuck you gotta do right here and now, or get outta my face."

All of a sudden, Prez and his mob wanted to be friends. Ranger knew that if he and Festus were going to get jumped and fucked up, it would have happened already. Why push it? Ranger and Festus strode back to their beers.

With everybody still in one piece, Ranger sat back down at the bar. He was pissed that the Squires MC had to deal with such stupid shit on a peaceful afternoon. Then he ordered a couple of hits of blackberry brandy to chase his beer.

Ranger simmered down and downed his drinks. Then one of the Mainliners came back over to the bar. He was staring Ranger down. He figured now was a good time to try and fuck with him. Ranger looked up from his drink, slammed both fists on the bar, and yelled, "Have *you* got a fucking problem, motherfucker . . . or *what*?"

The guy turned and walked back to his guys. Mel whispered to Ranger and Festus that maybe now might be a good time to split and head back home. But this thing wasn't over. There was no way in hell Ranger was leaving the bar. Nobody was going anywhere until *he* gave the word.

It would be a show of weakness on the part of the Squires MC if they split. Ranger ordered another round. Two hours later, the

Squires walked out the door, passed Prez and his boys, then rode out. Not a word was uttered between Ranger and the Mainliners.

On the ride home, Ranger figured he'd done the best he could. A good club president knows when to hold his mud, cut slack, and play the situation as best he can, for the benefit of his own neck as well as his club's. It was a decision Ranger had to make on the spot, one that affected the reputation of all his members. Although he would never openly admit it to Mel and Festus, he was glad it was over.

All in all, it was a good day. A hundred bike riders in the bar saw what went down, and it was bound to get around that the Squires stood up for themselves and gave a little push back. For the underdog Squires MC, it came down to the basics: Live to fight another day, fight to live another day, ride in peace.

Eli and the Raffle

This is a true story, so help me, and it involves a regular Joe named Eli. Now Eli always considered himself an unlucky cuss. He was forever digging himself out of one stupid fix after another. A couple of recent DWIs had left Eli sled less going on two years. If it weren't for bad luck, Eli would have no luck at all.

During the good old days, he'd owned his share of old beaters, starting with Jap dirt bikes as a kid, then his first Harley, a '58 XLCH. It was a chopper from a bygone era, a custom front end with a turning radius that took the whole damn street to flip a "U."

Eli tended bar at a place called Andy's Lounge, a wild little hole in the wall. Two doors down from Andy's was a 1%er clubhouse, which gave Andy's its edgy clientele, not to mention a perpetual line of scoots parked out front. And next door to the clubhouse was a tattoo parlor.

It was a slow afternoon at Andy's one day when VP Tony walked in from the clubhouse. Tony was good people. Back in the days when Eli still had a ride, he would sometimes join Tony and a couple of members on a short run or two as a hang-around. They even once asked him to prospect for them, but while he was flattered, it seemed safer for Eli not to affiliate himself—he rode with all kinds and poured for everybody else. Besides, it was Eli who always sent the drunken nasty babes juiced on Lounge Lizards down to the MC clubhouse for some hot action, which kept him in good standing.

That day when VP Tony popped into the bar, he was in fundraising mode for his club. He slapped a raffle ticket loudly on the bar.

"Hey, buddy," he yelled at Eli, "you're buying this. It's my last ticket, bro. One hundred bucks."

"Yeah, right," Eli said. But how could he say no to a 250-pound biker?

"No shit, it really is. First prize, a 2000 Dyna Super Glide."

Eli tore the ticket out of the book and threw it behind the register. Tony would have probably floated him the hundred, but Eli wasn't about to be indebted to a guy whose fists were bigger than Eli's head.

Eli wasn't much of a gambler. "Okay, pard. Here's your hundred," Eli said. As Tony stuffed the bill into the pocket of his cuts, Eli remembered the last time he threw a C-note worth of chips on a blackjack table in Reno and it disappeared just as fast.

That night after his shift, Eli dropped by the tattoo shop. A biker named Teardrop ran the place along with his old lady. As an artist, Teardrop was right up there with Buzzard and Filthy Phil. He'd been slinging ink for over thirty years. Teardrop began piercing thirty-five years earlier in L.A. His wife, a nurse, did all the piercing in the shop, but when it came to ink, Teardrop was the king. He wore a ponytail down to his ass and rode a black '57 Pan—a virtual showstopper—that always sat smack-dab in the middle of his shop like a museum piece. Teardrop had built it over the years with parts he'd bought and bartered doing his needlework. By now he had himself a bad custom scoot. Yeah, Teardrop was the man.

That night Eli and Teardrop got pretty hammered. Eli woke up the next morning facedown on the tattoo shop floor. Teardrop was out cold on one of his chairs. Shaking his head loose, Eli staggered out the shop door and headed back home on foot. By the time he got through his front door, either the phone or his head was ringing. Eli felt like death after a night of partying, but the caller ID on the phone told Eli it was 1%er Tony on the horn, so he picked up.

"Bro," said the voice, laughing on the other line, "come get your motorcycle out of my goddamn yard."

Eli nearly hung up the phone. He was in no mood or condition for games, but Tony was serious.

"You won the raffle, dummy."

The thought of winning the bike slapped him out of his stupor, and Eli fell on the floor and did the Curly Spin. Woo-woo-woo! The 2000 Dyna Glide was his! One hour later he was high on the hog, riding his ass off. It was like living a dream. A new scoot, Eli's first! Since he didn't have a garage, for the first month he parked the bike right in the hall of his flat. Every morning it would shock him. Somebody better come and fetch their motorcycle out of the hall. Then he'd remember . . . Hey! It was his!

Eli didn't have to do much to put his own stamp on the bike; just a few added touches did the trick. He changed the carburetor and installed a Screamin' Eagle system. He adjusted the exhaust system, added more chrome, and installed a backrest and a rack for easy traveling with his old lady. Riding became Eli's only escape, and the Dyna Glide became his ultimate escape vehicle. With the new bike, Eli finally felt as if his life—and his luck—was back on track.

If only he could say the same for his brother Teardrop. Over the next few months, TD's eyesight got worse. He began seeing double, losing his equilibrium, even going blind for a brief spell, which he blew off to too many hard years of partying. He thought he was going mental. Then the day Tear started hallucinating, he finally told his old lady, "Baby, I'm fucked up."

After a few tests and some MRIs, Teardrop was diagnosed with glioblastoma, one of the deadliest forms of brain tumor. The grim word from the local doctors was they'd give Teardrop maybe a couple months to live. Then they unplugged him and sent him home.

The hospital's treatment of their brother didn't sit right with Eli and his buddies. Everybody loved the charismatic tattoo artist. Jumping on his new bike, Eli delivered homemade Teardrop Freedom Fund buckets to every bike shop and barroom in the area to gather enough cash for a second opinion.

One of the brothers had an in with one of the doctors at the university medical center and scored a pre-op appointment with a high-priced specialist. Teardrop qualified for veterans' aid. But defeat was nearly snatched from the jaws of victory when all of the fifteen MRI images flew out of a box in the back of Teardrop's pickup on the way to the hospital. All of them were lost except for two images that were salvaged. But, as far-fetched as it sounds, Eli's luck still hadn't run out. The only two film sheets that were required to start the operation were the ones they managed to save.

The doctors removed 85 percent of the tumor, which meant that Teardrop didn't have to sell his '57 Panhead after all.

These days Teardrop's doing well. He wasn't pushed out to pasture by some two-bit HMO. And that's the story of how Eli's luck changed the day a raffle ticket was slapped on the bar. So remember that the next time your local MC needs your support for a raffle or a poker run. You may score a brand-new ride, and the brother's life you save may be your best friend's.

The Cannon
and the
Stakeout

The most common question I get
asked during interviews involves regrets: Looking back over four-plus
decades of bike riding with the Club,
do I have any regrets? I always tell them, "Yeah,
maybe three." First, I regret smoking cigarettes. Second, I took way too much cocaine.
Third, being a convicted felon, I regret losing
my right to own a gun.

I really miss the old gun runs. During our
weapon-collecting days, we sometimes walked
a fine line between which guns were legal and
which were considered illegal to own, especially considering our status as convicted
felons and indicted co-conspirators. Muskets,
pistols, and collector rifles would sometimes
fall into a gray area, although usually in our

cases, those gray areas tended to turn black once the Feds got involved. We rarely got the benefit of the doubt.

One time, we did.

A couple of our guys came across a Napoleonic-era cannon capable of firing a cannonball a little bit larger than a crochet ball. (Let's just say we scored it at a local flea market.) Restoring it became the perfect project for a wild bunch accustomed to dismantling and putting together cycles. (Let's just say our military experience came in handy, too.) We tore that baby apart, cleaned up the mechanism, and even soaked the wooden wheels in linseed oil. Pretty soon said cannon was glistening in the sun. One of the Club guys went down to the local gun shop, where the owner explained that muzzle-loading cannons were absolutely legal to possess in California.

Eureka! Now, could it still fire?

We took a broom handle, duct-taped a flashlight to it, and stuck it down the barrel. We were able to see light by looking down the primer hole. Sure enough, the piece was clear. Next we took a few ounces of gunpowder, pouched it in a small cheesecloth bag, and rammed it down the bore, chasing that with gasoline and a bunch of dirty rags from the garage. The guy at the gun shop specifically warned us: no way should we attempt to fire any kind of projectile. The cannon was that dangerous, and we took him at his word.

Good thing.

We aimed the cannon up the driveway between two houses, lit the fuse, and ran for it.

BOOOOOOM!

The wheels on the cannon jumped about a foot off the ground, spewing a curtain of blue smoke. The noise was sensational. Windows blew out of both houses. A woman from the house next door came running out from the back with a piece of her shorts in her hand. She had been on the toilet and jerked up her shorts so hard, all she had left of her drawers was a small piece of material.

After spending the rest of the day and most of the next with the glazier, we got all of the windows replaced and most of the neighbors settled back down. Still, we knew we could fire a lot more power and were dying to do it. We also knew the Feds had our houses under constant surveillance, and as hard as we looked, we could never figure out where the hell they were hiding.

Several of our ladies were taking a course in photography at the local junior college. For their final assignment, they needed to photograph an exciting action shot. Action shot? Hell, "action" was our middle name, right up our alley.

Just before sundown, we rolled the cannon across the street to another one of our houses. We loaded up and pointed it up toward the hill behind the house. This time we doubled the amount of powder, inserted even more rags, and poured what was left of the five-gallon can of gasoline down the barrel. The ladies were in position with their cameras. Ready? Then we lit the fuse.

BOOOOOOOOOOOOOM!

Smoke and flames leaped out of the barrel like we were fighting the Battle of Waterloo. In fact, the flames shot out so far, they lit the whole damned hillside afire, sparking up the darkening sky like it was broad daylight. We looked up the hill and out popped a whole mess of guys in camouflage clothing, jumping out of these bunkers dug deep under the ground.

Holy shit! It's the Feds!

As we stood there, mouths agape, they were running up the hill just as the fire department arrived on the scene. Thinking the Feds started the fire, the firemen were so pissed off, they hit the Feds with sticks and shovels and whatever else they could grab. We decided to take advantage of the confusion and jumped on our bikes and got the hell out of the area.

And the ladies? Snapping away during all the commotion, they turned in their action photo assignments on time. Well, of course, they all got A's—for Action.

Renton's Cool Sled

Renton had built himself one cool sled from what started out as a collection of everybody's spare parts. The '66 Shovel was an ex-PD auction bike, and when Renton laid out his wrenches, he didn't know a ratchet top from a star hub. But a bud from down south did Renton a huge favor when he risked life and limb and drove up to a local 1%er clubhouse, whipped out his camera, and snapped photos of some of the bitchin'-est rides parked out front. Renton was inspired when he got the photos in the mail. He blew up the snaps, tacked them up on the wall of his workshop, and got busy.

Renton's sled wasn't a chopper, mind you, but a traditional high-performance low rider, very similar to the FXRs that rolled out of the Harley showrooms years later, a bike ahead of its time. Renton kept the stock tank and the sixteen-inch rear rim. He raised the front end just enough to keep the bike level with a twenty-one-inch front rim. He added standard Avon tires, as much chrome as he could afford, a side mount taillight, a new plate mount, a Fat Bob rear fender, a Cobra seat, drag bars on short risers with nice rubber grips, lots of shiny levers on the controls, shorter drag pipes, a new swing arm mount, and then the icing on the cake: a super-clean-but-high-gloss black paint job with Oakland Orange pinstripes and flames (just like in the 1%er photos).

It was 11:00 on a Saturday night when Renton rolled out his custom scoot for its maiden voyage. The plan was to two-up with his old lady, hit the 405, then open up the Shovel in an LEO-free zone and go for a late-night spin. Although it was dark, the sky was typically overcast with rain clouds. As Renton strapped on his beanie helmet, he heard the crack of thunder. As he headed further north about three or four miles shy of the wide-open 405 interchange, the rain *really* started coming down in sheets. Renton slowed the Harley down to a dull roar and proceeded at the speed limit. Although the weather was for shit (pelting rain and windchill), fortunately the traffic on the four-lane expressway was light on both sides that night.

With no deference to the downpour, Renton noticed out of the left rearview mirror a strange pair of headlights barreling up in the fast lane, full speed ahead. Maybe it was a frenzied eighteen-wheeler making tracks in time to get home and out of the rain. Or better yet, maybe it was a ground-level UFO, or Big Joe driving his rig, Phantom 309. But as the vehicle drew closer, Renton saw that the lights didn't look quite right for a big rig; they were too low and too far apart. As quickly as those thoughts raced through Renton's mind, the lights approached closer and closer, faster and faster.

Renton pulled over to the slow lane to look over and see what the hell was moving so damned fast in this awful weather. He and his old lady saw two rigid-framed choppers pass them, black as the night. The bikes were nearly identical, long wide glide front ends, tall apes, no front fenders, and the loudest straight pipes Renton had ever heard.

Both men had very long hair and extremely long beards. They had no helmets on, were clad in leather from head to toe, and wore sunglasses in spite of the rain and the dark of night. Renton accelerated the Shovel to catch another look at the two 1%ers who roared past him.

A quarter mile down the road, under the gleam of a billboard light, Renton saw, strapped to both bikes' sissy bars, pump-action shotguns, butts down and barrels sticking up skyward. Just as Renton sped up to catch a better glimpse, both riders kicked ass and took off. They must have been traveling a buck and change as their taillights disappeared into the stormy night as quickly as they had first appeared.

Back at home, Renton couldn't strike the ominous night riders out of his head. Those two brothers were the real deal, all right. But where were they going? Or where had they been?

Renton picked up the local newspaper a couple days later. A story on the front page, just below the crease, caught his eye: NO SUSPECTS IN LATEST BIKER KILLINGS. The story revolved around an incident involving a biker who had been jumped by several other bike riders trying to cut off his patch. After taking a sound kicking and beating, the overmatched guy rolled over, pulled out his pistol, shot twice, and killed two of the assailants. Immediately after the shots were fired, the rest of the group pulled out their weapons and shot the victim full of holes, still lying on the ground. After the flurry of gunshots, everybody vanished. No witnesses reported seeing anything.

Renton contemplated what he had seen on the freeway. Then the newspaper story reported that two more bikers were found dead from shotgun blasts. Payback time, unnamed sources claimed in the paper. Renton finished his coffee, threw the paper in the trash bin, and kick-started his bike. Renton was a citizen, but if he was going to ride and look the outlaw part, there were certain rules and responsibilities that came with the game: Keep your head low, your mouth shut, and your business clean. Treat your brothers with fairness and respect. You don't want those two messengers of death coming after you.

Code of the Road

It's been said that showing up is half the battle. But when we're going on a run, showing up on time is a big piece of the puzzle when it comes to riding with the pack. Formation. Direction. Preparation. Organization. Coordination. It all has to be nailed down before we move an inch. Plus, serving time in the joint (not to mention the Army), I became sort of a stickler for time, especially about leaving. That tested some of the guys in the Club.

One time, the members started slacking off, not getting to the clubhouse on time to leave for functions. Finally, at one of our weekly meetings, I sort of lost my mind and went off.

"Look, when we say we are leaving to go somewhere at such and such a time, that does not mean you roll up at that time. That means you have to get here early enough so that we can get everything in order and leave together at the same time."

The following day, we were set to meet at the clubhouse at 8:00 P.M. to ride to Sacto for a party. Sunday evening, Cincinnati showed up early, and there was not a soul there. He decided to wait in the bar for the rest to file in. Comes eight o'clock, he was still the only one there, ready and waiting. He was plenty pissed, especially after yesterday's bitch-out, so he left an appropriately shitty note on the bar.

"Brothers. The pack left at 8, and it consisted of me. Signed, Cincinnati."

He got to Sacramento, and about an hour later he heard the rest of the bikes arriving. He came outside to meet the mob, standing there, feeling a bit smug, and noticed that no one was looking at him or talking to him.

"Ha," Cincinnati said to himself, "I think they all feel bad."

I got off my bike, glad that the rest of the guys held back their laughter. I walked past Cincinnati the Smug, and gave him a light smack on the head. "Daylight saving time, asshole."

Big Del is our pal and we've all been road-dogging for so long, it's unbelievable. He is one cool motherfucker, but he has his moments when he loses it . . . like the one time we were all heading cross-country, and somewhere in Utah everything went wrong. It started to rain. We all pulled over, and Delbert told me he had this special rain gear.

"So let's go. I have no problem with that, even though I don't have a stitch of rain gear myself."

Del suited up and we all took off. Twenty or so minutes on the road, I looked back and Del looks back at me with a big old grin on his face, happier than a clam. Then I look again and see the hood of his rain suit filling with air. Like a great big balloon—*bam!*—the hood blows up. Cincinnati was laughing so hard that he and the ditch almost got personal. Further down the road, we whipped into a little country store. Delbert was fairly dry, but his nifty rain gear was in tatters while Cincinnati and I were completely wet and muddy. So Del ran into the store and came out with a big roll of Saran Wrap. He starts making repairs on his rain gear by wrapping the stuff around his arms, legs, and giant body.

"Okay, let's go," Big Delbert yelled after his "wrap" session.

We weren't a mile or so down the road when I noticed these little bits of plastic coming off Delbert; he looked like a snowstorm. All of a sudden he hit the brakes and we came to a halt (again!). Delbert jumped off his bike like a madman, hopping up and down, pulling off the Saran Wrap, and trying to find an end. Now if you've ever tried to unwrap your lunch wrapped in that shit, you'll understand Delbert's dilemma, multiplied by one hundred.

We were laughing so hard we could hardly stand up. Delbert was swearing a blue streak. "Goddamn, goddamn, goddamn."

Delbert was soaking-ass wet from his impromptu sauna. Cincinnati grabbed his knife, leapt toward Delbert, and went to work. Cincinnati and I were still wet from the rain, not to mention the mud.

Heading into town to get a room, we split traffic by running down the white line. As Cincinnati turned his head, the wind grabbed his glasses and blew them clean off his face. Cincinnati is practically blind without his glasses, so we pulled over and Delbert walked back up the road, right through oncoming traffic, and found what remained of Cincinnati's specs. One lens was still serviceable; Delbert jerry-rigged the frames to sort of hang on his face. When we got to the motel, Delbert wondered aloud to Cincinnati, "It's a shame there's not some kind of operation or something they could do to your eyes."

We tell him, Yes there is. Delbert is amazed.

"You know, I'm gonna look into it when we get back home."

Back home, we'd forgotten about Cincinnati's wretched eyesight, but Delbert hadn't. He'd checked it all out: radial keratotomy, and Cincinnati was a perfect candidate for the procedure. Delbert was game and pushed Cincinnati.

"Let's go for it."

It's been a number of years since the operation, and Cincinnati's eyesight is tip-top. If God gave him a set of eyes that wasn't worth a fuck, then at least Delbert gave him a set that was perfect.

The man heard them a long time before he saw them. His first view was when he saw them coming down the off ramp from the interstate. He was an out-of-towner, parked in his car, in the shade cast by the gas station in Wall, South Dakota. Wall is located south of Sturgis, and the Sturgis Black Hills Run had started that very day.

He'd been watching groups of motorcycles all day, but this would be the first time he would be among them. He was not a motorcyclist himself. In fact, he did not know anybody who rode. The little he knew about cyclists was what he read in the newspapers and what he'd heard other people say.

In his mind, motorcyclists generally had bad reputations.

Eleven bikes (he counted them) pulled into the station like a swarm of locusts. They were all strung out, and even though there were a dozen or so of them, they took up more room than a dozen eighteen-wheelers. Some pulled up to the pumps, some stood in the doorway, while others lounged around the islands. Their presence was a picture of haphazardness, with no coordination at all. Eight of the motorcycles had a man and woman each on board. The others were just guys riding on their own. That made a total of nineteen people making enough noise for fifty.

Automobiles full of tourists started to form lines, waiting to get to a pump, as the cyclists were walking in and out of the store. How could the clerk possibly keep track of who purchased what, and who paid for whom? The women with the cyclists were all overweight. One was ghastly, enormous. She was clad in bib overalls and high lace-up moccasins of a terrible hue. She was the loudest and most obnoxious of the bunch, barking orders to everyone, especially the men.

Then it happened.

There were just three of them on motorcycles. They pulled into the station, took a look around, and rode over to the far side, near where the man was sitting in his car. They stepped off their cycles and one of them, surveying the situation, stepped up to the ghastly group of bikers at the gas station and asked, "Who's in charge here?"

The woman in the bib overalls pointed to a man indoors. When asked for his name, she produced it: Dino. The cyclist thanked her and walked indoors.

A few minutes later, Dino quickly emerged and hastily started to get things back in order. The noisy bikers shut their mouths and picked up their oil cans, candy wrappers, and soda containers littered around the pump islands.

The other rejoined his two friends, and when it was their turn at the pumps, they gassed up fast and were on their way.

The man, not easily impressed, nevertheless was.

These three men were members of the most notorious motorcycle club in the history of the world. And they knew how to keep order.

Heavily influenced by his older brother Big Ed, Irish and a small band of friends in a small community in California got involved with motorcycles. The small cadre of bike riders consisted of two brothers, Cool Breeze and DD, as well as Herc, Slo Mo, and, of course, Big Ed's younger bro, Irish. Big Ed felt that if they were going to get involved with bikes, then it was up to him to at least pound into their heads not only biker basics, but the basic principles of life. As friends, they didn't really need too much schooling, because they were an honorable lot; it was just that they were really young.

Nevertheless, Big Ed invested the time and was very proud of them.

After the guys had been riding for a couple of years, Big Ed was killed in a huge wreck, but his influence was felt long after his death. One evening the five young guys sat down and decided to map out their destiny. Seeing as they had complete and utter trust and respect for one another, they decided to form a club. After much discussion and debate, they agreed that the decided purpose of the club would be to eventually form their own charter and then be absorbed by the leading 1%er club in the state. They would attack this goal with much vigor.

It took them a while, but with pride, honor, and perseverance, their goal was finally attained. They had been asked to join, en masse, with this leading 1%er club. Their dreams would soon be fulfilled.

There was just one problem. There had been a violent incident with another group, and even though they were working to resolve it, it had remained ultimately unsettled. The problem was that they did not want to take this bit of unfinished business with them to the 1%er club because it did not involve the club they were advancing into, and they damn sure were not going to let it pass. If they turned down the offer to be absorbed, it could be misunderstood and never again be offered.

Cool Breeze stepped forward with a plan.

"Look," he said, "you four guys go ahead and join. I will remain behind as a member of our old club and take care of this on my own."

The four were immediately up in arms, but Cool Breeze would hear no argument.

"I know as well as any of you that if this had been your idea, you would be offering the same proposal."

It was agreed that Cool Breeze would stay behind as the lone member of the old club, and that upon completion of their vendetta, he would completely disband the old club, burn the last patch, and join his four brothers, wherever they might be.

"Don't worry," warned Cool Breeze, "if you don't hear from me, there's no sense in making contact until the job is done."

The friends all agreed.

It wasn't long until the offending club heard the news, knew that they had fucked with the wrong guy, and scattered to the four winds. But Cool Breeze was patient. All he had was time and a list of names. He knew what had to be done.

It was just about three years later, and a bike was heard one night pulling up to the new clubhouse. There was a knock at the door. Somehow the members already knew who it was. A little older and stronger, and a lot wiser, Cool Breeze rejoined his brothers. He became a prospect and then a member. A debt of honor had been paid.

The Worcester Mass

On a brisk Halloween Day, 2002, in Worcester, Massachusetts—thirty-eight degrees outside—Big Scott stepped up to the microphone. He was set to deliver a tribute he'd typed out the night before, one he had dreaded delivering all of his life. Lately it seemed as if the highway were lashing back. Big Scott had lost another bike-riding brother on the road.

Nelly was the most famous casualty. He was killed in 2000 in Ipswich on Route 133 when a nineteen-year-old driver crossed into his lane. Nelly's publicized death sent shock waves through a large bike-riding contingent throughout the state. It inspired the Massachusetts legislature to pass a bill that would soon require a motorcycle awareness video as part of the driver's education process. The annual Nelly's Run in Cape Ann continues to draw over a thousand riders every August to help raise scholarship money not only for the two kids Nelly left behind but for other children of motorcycle fatalities.

In August of 2002, Carlos went down. Carlos was a giant man, six feet four inches, three hundred twenty-five pounds, and a self-employed truck driver who wore a full Santa-type beard. Carlos keeled over from a fatal heart attack on his Harley while partying on the road at Daytona Bike Week.

Dying and riding had become too damned synonymous in New England. Big Scott lost a host of his buddies on the roadways, but none hit as close to home as the man they were about to bury on Halloween.

As a young man in the 1960s, Gomer rode the streets of Worcester on his motorcycle. As a child, Little Scott hung around his dad's electric shop every Saturday. Those were the days before they built the large industrial parks on the outskirts of town. Back then, industry clustered locally around Park Avenue and Main and Chandler Streets in Worcester.

Every Saturday at noon, there was a mad dash for the exits at the electric motor shop as the workers who had put in their weekend overtime charged for the doors. Scott and his dad would head out to the parking lot, and after the last employee would drive off, they would close and lock the gate. It became a father-and-son ritual.

Then Little Scott first laid his eyes on Gomer's ride.

"I saw this guy, sitting on a motorcycle. I was so young I had never seen anybody start up a bike. What was this guy doing?"

Gomer kick-started his Harley Panhead. After three tries, the Pan exploded. Huge chopper. Straight and trumpet pipes. Extended front end. Big sissy bar. The bike bellowed and roared. In one fluid motion, Gomer jumped on the cycle, hiked the kickstand up, swooped down, grabbed his helmet, and slapped it on his head. He didn't bother tying the strap. He gunned the motor, shifted into second, and headed straight for Little Scott.

"I don't know if he saw me or even gave a shit. Just then my father pulled me up out of the way by my right arm. My feet were way up off the ground. I was flapping like a banner in the wind as this madman on metal came blasting by."

In seconds Gomer was gone, darting in and out of bumper-to-bumper traffic. Little Scott looked up at his dad, who locked the gate as if nothing unusual had happened.

"Someday," he told his father, "I'll ride one of those."

Little Scott's dad gave his son a cuff across the head.

"Your mother's not gonna like that."

Gomer was nineteen, straight out of trade school, when he started working for Scott's dad at the electric shop. He was ready to take on the whole world. He loved the ladies and juggled two or three girlfriends at a time. But he loved his motorcycle even more. It was 1969, a good time to raise hell. Gomer just missed getting shipped off to Vietnam. He was eligible for selective service and never went to college, but somehow he avoided getting drafted. Nobody knew quite how.

Still a kid, Scott kept close tabs on the motorcycle parked in the lot at the electric shop. He noticed even the slightest changes: a chrome part, a new tire, or a dipstick with a gauge on it. Scott's first ride on a motorcycle was on the back of Gomer's 1972 XLCH. When it got stolen, Gomer bought himself an updated '74. The '74 Harley eventually became Scott's first bike, an AMF suicide machine that stayed in "the family" and was handed down to various buddies and friends for more than twenty-five years. It had no front fender, so when it rained the water splashed Scott square in the face as he rode. Occasionally, the front headlight would short out, and since there were no front brakes, the bike looked ultra cool. Riding the '74 in the rain, Scott dodged tractor-trailers and learned how to glide, ride, slide, and survive.

During the late sixties and throughout the seventies, Gomer used to ride up to New Hampshire for Laconia Bike Week. It was a wild, bygone era for bike riders. When Scott hit age eighteen in 1978, he soldiered his first trip to Bike Week with Gomer. It was the year the two camped by the side of the road in the pouring rain. It rained so hard that some 1%ers took pity on the two soaked bike riders.

"Hey, assholes, here's a coupla trash bags to sleep in, so you'll at least keep dry."

Once the two rolled into Laconia, Gomer and Scott hit Cascade Park, a trailer park on Route 106. Cascade Park was down the street from the main campground, aptly named Animal Kingdom. During Bike Week, Animal Kingdom became a ramshackle, motorcycles-only racetrack. Men rode radical, borderline-illegal choppers, had real long hair and beards, wore tattoos, and flew three-piece patches. You could buy a beer, sausage, club support T-shirts, or a set of cool leather saddlebags for cheap.

Since a rough throng of 1%ers dominated Animal Kingdom, camping there meant no sleep. So Scott and Gomer doubled back the five miles on Route 106 and slept at Cascade Park. During Bike Week, Animal Kingdom and Cascade Park were reserved solely for bike riders. Tents were pitched and a couple of overflowed outhouses serviced the whole area. At sundown, riders clutched their bottles of Jack Daniel's, partied, exchanged stories, and talked motorcycles all night long. By sunup, riders would amble over for a pancake breakfast and consume piles of starchy flapjacks. Then it was back to the Animal Kingdom to watch more bike races and hang out. On the way, Gomer and Scott saw a lynched Jap bike hoisted up in a tree and set on fire. Topless biker babes strolled nonchalantly along the roadside.

Gomer got a biker cartoon tattooed on his shoulder by Tattoo Tony, who was half in the bag. Tony was so drunk, the tat had to be touched up a couple of months later. Gomer got tattooed in New Hampshire because up until 2001, parlors were illegal in Massachusetts.

Meanwhile, a rider named Flyfisher walked around Laconia, wearing nothing but his chaps, his bare ass hanging out. He had ridden his motorcycle down Route 3 by Weir's Beach when he got arrested. In court, he stood in front of a judge, still wearing nothing but his chaps.

"Son, put your pants on, pay your fine, and don't come back," the judge declared. Flyfisher was a decorated Vietnam vet. After a few hitches in 'Nam, he had seen everything. Back then it was all about raising hell.

After Laconia, Gomer bought a '77 Low Rider. It was an ugly bike, gray and black with a red stick-on Harley insignia. The bike religiously leaked oil. Scott and Gomer rode together to numerous bike hangouts, places like the Bach Dor Pub on Route 6 in Chaplin, Connecticut, where every Sunday during the summer the bar hosted a massive barbecue and pig roast. Big Jay was there, and Carlos, too. Tattoo Tony's bus was parked nearby. He'd set up shop to sling ink in the parking lot near the pub.

One night, back in Worcester, Scott got a phone call. It was Gomer. His bike was broken down. Again. It was past midnight and he was fifty miles from nowhere on a rainy night. Scott dragged himself out of bed and borrowed a pickup truck from the electric shop. He searched out the disabled Harley and found the bike and Gomer in the company of two attractive women who had just fed him a steak dinner and way too much wine. Scott struggled to wheel the dead bike up into the bed of the pickup while Gomer had one of the girls on the hood of the truck with her shirt open. Back home Gomer insisted on unloading the bike himself. Half drunk, he backed the bike down the board ramp without any brakes, jackknifed the thing into an embankment, and ended up face first in the mud with the bike on top of him. Scott finally poured him into his bed, muddy and filthy. Gomer phoned the next day.

"Thanks for coming out to get me."

Gomer's last ride was his return trip from Biketoberfest 2002 in Florida. On the way back to Worcester, he and a couple of riders stopped off in Holly Ridge, North Carolina. The riders pulled off to the side to stretch their legs. Gomer turned around and headed toward a restaurant they had passed a few miles back. While circling back onto the highway, Gomer didn't look twice. A sixty-one-year-old man driving a Toyota Camry struck him. According to the local police chief, Gomer collided head-on with the Camry, then bounced on the hood and struck the windshield. His Road King landed almost five car lengths away from the Toyota. Gomer died in less than twenty minutes. No charges were filed since it was determined that the driver of the automobile had not been speeding. Gomer left behind a mother, brother, niece, and nephew in Worcester and a slightly banged-up Road King back in North Carolina.

Once again Big Scott got the call. Days before the funeral, Scott made the 1,700-mile round-trip to North Carolina and picked up Gomer's motorcycle for the last time. It was a long, tiresome journey, but Gomer's bike and his body were finally back home in Worcester where they belonged.

At the funeral, patch-wearing pallbearers from rival clubs set aside their differences and laid Gomer to rest. Big Scott spoke his peace.

"Gomer loved good times, women, honest work, a good joke, motorcycles, family, friends, and food. He knew more about Harley-Davidson motorcycles than anyone I've ever met."

About a hundred bike riders made it to the funeral and caravanned to the gravesite in the Massachusetts chill. After the service, most agreed that Gomer would have loved having his funeral on Halloween.

The following Friday, Gomer's death still hadn't sunk in, so in his honor Big Scott rode solo up to Laconia. His birch-white 1984 FXRP, a former police bike, hummed beautifully. The glide of the machine on that golden, sunny November day was flawless. Big Scott negotiated every curve and straightaway in his memory. It was as if Gomer were still there, riding next to him. He admired the magnificent New England landscape and remembered a bumper sticker he had on his desk back home. When he made it back, as a tribute to his late friend, he slapped the sticker on the back of his truck. It read: "Check Twice. Save a Life. Motorcycles Are Everywhere."

Active Development

Like lots of Southern Californians, Jules had had his fill of the Hollywood life. On paper, it all looked great. He was making headway as a budding screenwriter, though lately he'd hit a creative stone wall of sorts. Jules came to sunny Hollywood from snowy northern Michigan on his Harley Fat Boy with a leather satchel full of what he thought were catchy ideas, spicy pitches, and final drafts. However, over the past few years, Jules had spent most of his time untangling the words of others.

His first real writing job was on the Paramount lot rewriting and cleaning up sitcom scripts. The last year and a half he'd been a script doctor, doing movie rewrites. His last project was an action picture based on a popular comic strip. Writing for films was all about achieving a screen credit for his work. To Jules's disappointment, at premiere after premiere, his name was nowhere to be found on the screen. But he'd expected it. It was little consolation that one paycheck for a script could keep him going for six months. What mattered most to Jules was whether or not he continued to have a reason to write and ride his motorcycle.

On the car ride back home from another screening, there was a disappointed silence between Jules and his lovely live-in girlfriend, Tania. Her career hadn't gone far. A few modeling jobs at a couple of conventions in Long Beach and small parts in a repertory playhouse in Pasadena. Jules knew she was counting on him to pull them both through, which only added pressure to their relationship.

It used to be that a scoot on the Fat Boy was all Jules needed to rejuvenate the juices. A quick spin along the Santa Monica piers would blow some of the mental cobwebs loose. The bike's velocity helped restore his confidence on the page. An hour later he would be back at the word processor, turning phrases, mumbling dialogue to himself, and formulating plot twists and surprise endings.

But lately, even a ride on the Fat Boy came up dry in replenishing Jules's confidence and inspiration.

That's when Jules decided on a radical change of scenery. He could write anywhere. Maybe he didn't need to live in the daily crosshairs of the movie business. Besides, the people could be so false. It was the land of "the deal," and unless you had one, no matter how big, no matter how small, you were part of the underclass.

It was summer in Santa Monica, although that didn't really mean much. This was the land of perpetual sunshine. Jules had reached an impasse on his current script, one he was writing for himself. He pictured it as a small independent film, perhaps playing at one of the local art house screens in town. It was a coming-of-age story. A young Iowa Hawkeyes fan moves to Michigan with his family. His tyrannical father was an auto executive. It was sort of autobiographical, mostly not.

Right around the second act, Jules and the dialogue ran out of steam. He pushed himself away from his keyboard and reached for the keys to the Fat Boy. It only took ten minutes before he found himself sitting on a park bench along Ocean Avenue, staring at the pounding surf of the Pacific Ocean, the Fat Boy parked assuredly by his side. And that's when Jules decided.

He would definitely leave L.A. on his bike.

Over the next three weeks, he cleared out whatever was cluttering up his life. He broke up with Tania. He placed three weekly ads in the local freebie paper, advertising his garage sales. Each weekend, he sat stationed on his lounge chair, munching Krispy Kremes and watching a variety of curious customers rummage through the material remnants of his life. Only two possessions mattered in his life now, the Fat Boy and a mint 1966 Bultaco Astro 250 dirt bike that he'd acquired as a gift to himself when he got his first movie-writing job. While the Fat Boy stood sentry over the garage sale, the Bultaco was stashed deep in the garage. Jules was iffy about selling the bike until a familiar face walked up the driveway and asked about the 250.

"So, how much for the Bultaco?"

"Oh, yes, the Spanish bike." The face was familiar, so familiar that Jules was taken aback. What was *he* doing casing garage sales?

The dark-haired actor, ruggedly handsome, put his forefinger to his lips, as if to shush Jules, anxious to maintain his cover. Neither of Jules's two other customers, the Russian or the Korean bargain hunter, recognized the famous figure. The actor had just broken things off with his wife, a noted actress twice nominated for an Oscar, and lately both had been getting steadily trashed in the tabloids. The paparazzi depicted the actor as heavy, hotheaded, and hairy, but in the flesh, in the dim light of Jules's garage, he was buff, fit, and personable.

"She's a beauty, all right," Jules said to the actor. "It hurts me to sell her."

The actor flashed him a toothsome grin, then pulled out his checkbook.

"Yeah, but think of it this way, you'll be giving her a good home."

For someone so wealthy, the actor drove a hard bargain. The two settled on $2,500. The check surprised Jules. He figured a man of such stature would carry at least that amount of cash on his person. When Jules exchanged the pink slip for the actor's personal check, he had the fleeting absurd notion of not cashing the check, as if someday the value of the signature might eclipse the Bultaco. Then he came to his senses, hit the ATM, and deposited the money on his way to catching a bite at Chez Jay on Ocean Avenue.

The three garage sales had been a raging success. What furniture, apparel, and knickknacks Jules wasn't able to sell, the Salvation Army happily carted away. Only his Fat Boy, a week's worth of clothes, and approximately $17,000 that he'd converted to traveler's checks and hard cash remained of his Los Angeles lifestyle. He was ready to move on. The lease on his bungalow would expire in a few weeks. Tania moved her stuff out. Leaving now meant anytime he felt the urge.

The following Tuesday, the urge came. He stuffed his saddlebags with what little clothing and toiletries he needed. Jules would head north on Interstate 5, then hit Highway 101. Destination Mendocino. Find a small shack. Resume his writing from a fresh perspective and a new environment. Get closer to nature. Ride his motorcycle whenever he pleased, free of pollution and congestion. If anything popped, he'd merely be hours from an airport, or even better, no more than a couple days' ride to Los Angeles to "take a meeting."

"This is perfect," Jules said to himself as he closed the locked door of his bungalow and, as promised, tossed the keys through the front door mail slot. Then it was good-bye, L.A. He would have one last meal at Chez Jay, then head off toward the 405 and hit the Grapevine on 5.

It was midway between lunch and dinner when Jules hit the door at Jay's. The flood of sunlight lit even the darkened corners. As he'd suspected, the place was empty. Except for a lone woman sitting at the bar. Jules usually ate at the bar, so he situated himself not far from where the lady sat. She was an obvious beauty, an actress type. Her blond hair had been brutally straightened. Her face was heavily made up. Except for her casual attire—jeans and a blue-and-white-striped pullover shirt—the straightness of her hair made her look like, well, just like Veronica Lake. A leather jacket was slung over her chair.

Jules took a mental photo of her face. She was an L.A. classic, part of the reason he'd hopped on the Fat Boy to come to L.A. in the first place, part of the reason he was now leaving. Somewhere down the line she'd make a good character. The two mumbled pleasantries.

"How ya doin'?" Jules nodded at the woman.

"Been better." She repeated the phrase, the second time under her breath. "Been better."

When she spoke, she should have been puffing on a cigarette; her hickory voice was smoky and sexy. For a moment, the tone of her voice transported Jules back to the 1940s. He was having one of those film noir moments.

After a pause, she seemed to perk up.

"Nice motorcycle," the woman remarked as she pulled her mixed drink away from her lips.

"Thank you. Right now she's in her prime. Just got her tuned up and serviced. Runs like a watch."

The waitress brought Jules's regular order. Pint of lager, Foster's. Next up would be a roast beef special. This would be the something Jules would miss about L.A. The food. And the women. He paused again to admire the woman's beautiful profile.

"Don't suppose you ride?" he asked.

Her answer surprised him.

"As a matter of fact, I do, though not for a while." Jules noticed that she spoke in a slight, faded British accent.

"Pray tell."

The woman turned her face toward him. For the third time in a row, he was taken aback. Her cheekbones were high, her nose was aristocratic, and her skin looked as soft as cream and white as porcelain.

"My husband is a director. We have a garage full of Harleys, a few Indians, too. I learned to ride primarily to keep up. I figured that would give me the edge."

"The edge?"

"Over the women, you know, the actresses, models, the beautiful people. Occupational hazard in that world, you know."

Jules shook his head. Yes, he knew about the occupational hazards of the entertainment business. Then the woman extended her hand and introduced herself.

"Molly."

"Pleased to meet you. I'm Jules."

"Jules. Nice name. What's your story, Jules? I told you mine."

"Mine's much simpler. I'm leaving L.A. I'm a writer who has just decided I don't need to be here. I can write anywhere. What you see outside, that bike, that's the extent of my life. I'm cutting my ties and taking my chances. Maybe I'll land on my feet. Maybe I'll fall on my ass. All I know is that I've got to find out if I'm worth a dime away from here. I can't keep living like this. My life is in 'active development.' It needs to be 'green lit.'"

Molly laughed at Jules's command of the jargon and shook her head knowingly.

"We all dream," said Molly as she sipped her drink. She spoke as if she were reciting her lines. "We dream about getting away. Escaping. Leaving behind our fettered routines. But few of us have the balls, or the ovaries, whichever applies, to actually do it."

Jules snorted a laugh. "Well, I guess I have the balls or else the lack of good sense, because, hell, I'm doing it."

It was just a hunch, but the light from her eyes and the radiance of the smile on her face told him it had been a while since a grin had crossed her lips. Whatever was ruminating in her mind, it was making Molly more and more appealing to Jules by the minute.

Just then the waitress brought over Jules's plate. Nothing beat the meat and potatoes at Jay's. Jules looked over at Molly.

"Excuse me while I eat my final L.A. meal?"

"By all means." She smiled, raising her glass. "Here's to you."

Jules shoveled his food like there were no tomorrow. Tender shards of beef he could cut with a fork. Red potatoes. A small Yorkshire pudding. It was all gone in minutes.

"My mother used to make Yorkshire pud every Sunday when I was a little girl." Her voice trailed off and her eyes grew glassy.

Jules looked up from his empty plate and stared at Molly, who was gazing back at him, then at the Fat Boy, then back at him. She reached over and touched his elbow.

"Take me with you."

T**he roar of the Fat Boy drowned out any possibility of com**munication. But Molly didn't have to say a thing. Jules knew what she was thinking. He didn't have an extra helmet, so he offered his to Molly. When she demurred, something about risking it all, Jules decided to keep it stashed. He'd take his chances with the law and feel the wind. Molly lowered the foot pegs and tied her hair back. Fortunately she'd worn a leather jacket, unusual wisdom for a late summer Southern California afternoon. She wrapped her arms around Jules. A babe on the back of his Harley. It was a feeling he never got used to, nor took for granted.

Jules and Molly gassed up near the Magic Mountain amusement park. As Jules walked toward the station's pay counter, Molly showed her prowess with a gas pump. By the time Jules made it back to the bike, they were filled up and ready to resume the ride.

The whipping winds on the Grapevine tried their best to beat back Jules and Molly, but their wills, bolstered by the torque and power of the Fat Boy, proved stronger. The Angeles National Forest looked more like a scorched mountain hillside. But Jules was feeling free, and Southern California looked even better as it got smaller in his rearview mirror. For Molly, it could have been Scotland or Wales for all she cared. She lay her head close to Jules's head, on his shoulder, and maintained a desperate grip on him.

It was beginning to grow dark. Jules decided at the 120-mile mark that it was time to call it a night. He'd covered enough distance to consider himself on the road and outta L.A. By the time they hit Buttonwillow, Jules noticed a cluster of motels, fast food joints, and truck stops. Next stop, Big 6 Motel. So as not to be presumptuous, he'd book a pair of rooms and then plan their best course of action, an early start the next morning toward Mendocino.

Jules killed the bike in front of the Big 6 office. He and Molly dismounted and walked inside to book the rooms.

"So," Jules said to Molly. "Poolside or what? I like mine facing away from the highway."

"One room or two?" the Indian clerk deadpanned.

"One, please," Molly intervened.

"Then it's on me," Jules said, pulling his money clip out of his pocket. "One night in advance, please."

"Two beds or one?"

"Whatever," Molly answered.

Jules felt a little awkward that Molly didn't have a lick of luggage, only her leather jacket. After the manager presented a card key, Jules and Molly putted the bike the short distance to the room located on the far end of the motel complex, away from the pool. Jules parked the Fat Boy in front of the window, where it remained in plain sight all night.

Room 68 was stuffy and hot, so Jules cranked the air conditioner full blast. Jules pointed to the phone next to the king-sized bed.

"Don't you need to make a call or something?"

"I don't think that'll be necessary."

"Hungry?"

"Not really," she responded with a half-sigh.

Jules sat on the bed. Molly sat next to him, closely. Like she had on the bike, she perched her head on his shoulder. Her demeanor was wistful as opposed to passionate.

"Why don't we just relax on the bed?" Jules suggested. "Those winds tend to tire you out and sap your energy. Besides, my arms ache a little bit and we have a long stretch to cover tomorrow."

Molly smiled faintly and shook her head agreeably. With the same lack of self-consciousness many actresses have, Molly pulled off her shirt. Underneath she wore a thin blue tank top with no bra. Then she kicked off the penny loafers she'd been wearing and brazenly stepped out of her jeans. Her legs were thin and long. Jules followed suit, stripping down to his boxers and Paramount softball team T-shirt.

Jules positioned himself on the middle of the bed, protectively wrapping his arms around Molly. He'd only known this woman a matter of hours and already he felt the need to look after her. Not long after her head hit his shoulder, she was quietly sleeping. Jules spent the next few minutes admiring their reflection in the motel bureau mirror. The clock radio read 8:30, early evening, as Jules felt a lingering sadness wash over him, as if Molly's reserved sorrow were contagious. Not long after, he dozed off as well.

I t was midnight when Jules awoke to the sound of the television. Molly's head still lay on Jules's shoulder as she toyed with the remote control. Jay Leno was on the tube, bantering about the Hollywood elite with a second-generation actress trying to waltz out of the shadow of her famous mother. A reproduction of a Los Angeles skyline sparkled behind Jay's coiffed hairdo. Oddly, a Harley-Davidson was parked in front of the desk as Leno and the female guest tittered about the possibility of riding away together. The actress giggled. Jay comically shrugged his shoulders. Then a vague look of longing crossed Leno's face.

"Hope I didn't wake you," Molly whispered.

"L.A.—there's no escaping it."

Without warning, Molly placed her lips on Jules's and then rolled over on top of him. After a few deep kisses, he felt her tugging at his boxers. She peeled off her panties and straddled him. Extending her arms toward the cottage-cheese ceiling, Molly pulled off the blue tank top. In the light of Leno, it was like watching a panther stretching. Sleek and strong. She placed her palms on Jules's chest and he entered her. He cupped his hands over her breasts and held on as she grinded her hips and rocked her body in a slow circular motion.

It all happened pretty quickly, more rapidly than Jules was used to. Both huffed and puffed for a few minutes until Jules came rapidly, and he was pretty sure she hadn't. Then the two lay close again, arms entangled. Jules felt almost apologetic for his poor performance. Molly didn't seem to mind, as if she preferred the spontaneity to the act itself.

"I'd kill for a cigarette," she said to the sound of studio audience laughter. It was a welcome sound Jules knew well from his sitcom days.

"Sorry, I don't smoke."

"Don't be."

Jules looked down at Molly and noticed the natural curl coming back to her dirty-blond hair.

"I've smoked all my life," she said. "At this point, it's kind of silly to quit."

"Dangerous stuff," Jules concurred.

"Better to live dangerously than cautiously."

"Meaning?"

"Meaning nothing," she said. "All that talk about ticking time, it's true. That's what I like about motorcycles. It slows down time, at least to a manageable level. How I've missed riding a bike. Our little trip woke up my senses. That ride over the Grapevine was the most alive I've felt in months."

A strange feeling swept over Jules. Suddenly he missed the brilliant night-lights of Southern California. Those Hollywood nights, like the song said. And Tania. And Chez Jay's. The pier. It was a longing he hadn't expected.

"Ever think about dying?" Molly asked Jules.

It was a morose question, and frankly, immortality didn't cross Jules's mind very often. He was one of those live-day-by-day kind of guys.

"Not really," Jules said, "How about you?"

Molly didn't answer.

The next morning it was obvious that Jules and Molly were headed in opposite directions. By morning, Jules had decided to head back to L.A. He could feel the magnetic pull of show business tugging at him, demanding his return. He was on the brink. He was in active development, on the verge of being green-lit. Jules broke the news to Molly over a Grand Slam breakfast at the local Denny's. She didn't seem surprised, then ordered a large glass of orange juice.

"It's an addiction, I suppose," she said. "You'll get over it."

"What will you do?" Jules asked her.

"I intend to move on. I'm going north, for a while. Then south. Then east. Then north again. Then we'll see."

Molly paid the tab for the Grand Slam and the OJ. Then the two walked out to the windy parking lot, then over to the Fat Boy. Jules impulsively threw Molly the keys.

"Here. You ride. Take her."

As if anticipating both the toss of the keys and Jules's mad, generous act, Molly speared the key ring. She offered no polite resistance.

"You'll need this as well," Jules said, pulling the pink slip out from his wallet and scrawling his signature on it.

"I'll have the necessary funds mailed to you."

"Take your time. The address is on the slip."

"Thanks, ah—" she checked the slip for his name—"Jules. Jules Raymond. I'll never forget this."

Molly mounted the Fat Boy and with a commanding confidence started it up and revved the throttle. She offered no parting enlightenment except a smile, not even a good-bye kiss. He watched her ride off, satisfied that she'd more than make it north, wherever she was bound. After Molly rode out of sight, Jules pulled out a hundred-dollar bill and walked over to where a band of south-bound truckers had parked, hoping to at least grab a ride to Magic Mountain, where he and Molly had originally gassed up.

Two months later, Jules was typing away, fixing another action script, when he heard a knock at the door. When he opened it, a tall, thin man in his early fifties stood at the door. He spoke in a reserved southern drawl. He could have been an Ivy League history professor; his graying beard was trimmed neatly. He held out an envelope to Jules. On the corner was the name and logo of a prominent Hollywood production company printed in art deco script. It was a film house Jules would have died to write for.

"Jules Raymond?"

"Yes."

"Do you mind if I come in?"

The house was bare. Jules was ready to apologize for the starkness of the rooms when the man quickly walked past him and into the dining area. He extended his hand.

"I'm Jeremy. I believe you were friends with my wife." He handed Jules the envelope.

"Open it," the man said.

Jules tore it open. It contained a check for $25,000, more than fair value for the Fat Boy.

"Listen, sir, I can explain."

"Mr. Raymond, there's really no need. I'm indebted to you for far more than what your bike is worth. Because of you, my wife's last days were filled with adventure, excitement, spontaneity, and happiness. More than I had time to give her, in fact. If it's all right with you, I've added your bike to my collection."

"I don't understand."

"I'm sorry to tell you this—I don't know how close you two were."

"I only knew her a day."

"I understand. We buried her last week. Female complications, you know."

"Actually I didn't." Jules paused to remember her face. It was a pleasant memory. "She sure knew how to handle a bike, Jeremy."

"She did. Absolutely."

Jules looked down at the check, then back at the director's tanned face. The man looked resigned, though hardly distraught or grief-stricken. Perhaps he had been an actor in his previous profession. Maybe he had mastered the art of composure. After a few minutes of small talk, the man left the house in a hurry and drove away in an expensive Porsche.

Walking back to his austere writing room, Jules pushpinned the check to the bare wall and studied the signature for a minute before returning to his screenplay. Four months after, Jules still hadn't cashed the check.

* * *

I f you enjoyed "Active Development," understand that it is entirely a work of Sonny Barger fiction. And there's more where that came from. In October 2003, Sonny will publish his first novel, *Dead in 5 Heartbeats*. In it, Sonny will introduce his fans to "Patch" Kinkade, former president of the most powerful motorcycle club in California, the Infidelz. It took a lifetime of riding the highways, cracking heads, and sparring with John Law for Sonny to create a character like Patch, a tough-as-nails warrior whose primary concerns are loyalty to his brothers, respect for his machine, and love of the open road.

Closing Credits, Contributors, and Unindicted Coconspirators

Thanks to the following people who provided inspiration with their hell-raising motorcycle tales:

Andy Ubell

Barbara McQueen Brunsvold

Brenton Demko

Dave Blake

Dave Daugherty

Debbie Tolly

Dirk "Butch" Walter

Don Whitney

Ed "Animal" Cargill

Guy Girratono

James Fitzpatrick

Jeremy Povenmire

Joe Payton

Kaye Cheshire

Linda Black

Marjorie Goodness

Mark "Chilli" Brehm

Marvin Gilbert

Michael Gervais

Mike Locarro

Randell Widner

Richard Charles Anderson

Robert Jypson Woods

Robert Scott Edwards

Ron Braithwaite

Ronald French

Sara Laliberte

Sherry Anderson

Teresa Midkiff

Tobie Gene and the East Bay
    Dragons MC

Scott "Big" Bigelow

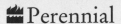 Perennial

Books by Ralph "Sonny" Barger:

DEAD IN 5 HEARTBEATS: *A Novel*
ISBN 0-06-053251-3 (Coming Fall 2003 in hardcover from William Morrow)

"Patch" Kinkade, the former president of a powerful motorcycle club in Northern
California, is hoping to start a new life in Arizona. But when bad blood between
members of rival clubs litters a casino with the corpses of both club members and
ordinary citizens, Patch slips on his leathers, straps on his knives, wipes the dust off
his Harley, and cruises down the highway for what could be his final ride.

RIDIN' HIGH, LIVIN' FREE
Hell-Raising Motorcycle Stories
ISBN 0-06-000603-X (paperback)
ISBN 0-06-009522-9 (audio cassettes) • ISBN 0-06-009523-7 (audio CDs)

Rousing, moving, wildly entertaining true-life stories of Ralph "Sonny" Barger's
renegade brothers and sisters in black leather and their relentless pursuits of liberty,
individuality, and the "ultimate ride."

**"A peek at another side of America from an interesting personality who lived
it and still revels in it."** —*Tulsa World*

HELL'S ANGEL
The Life and Times of Sonny Barger and the Hell's Angels Motorcycle Club
ISBN 0-06-093754-8 (paperback)

The Hell's Angels have been the most notorious group of motorcycle bad boys
in America for the last forty years. Perhaps the baddest Angel of all is club
visionary Sonny Barger—who has been sanctioned by the club to tell the truth
about the tight-knit group of free spirits who are simultaneously envied and feared
throughout the world.

www.SonnyBarger.com